D0365184

In this series:

H91.57 . M37 P5 1981

oclc + nooalinc
6-4 2010

Mathematical models in biological oceanography

Edited by T. Platt, K. H. Mann and R. E. Ulanowicz

Walford Library
U.S. Dept. of Commerce, NOAA,NMFS
Northeast Fisheries Science Center
74 Magruder Rd.
Highlands, NJ 07732

The Unesco Press

39298100000997

Published in 1981 by
the United Nations Educational, Scientific
and Cultural Organization,
7 Place de Fontenoy, 75700 Paris
Filmset by Asco Trade Typesetting Ltd, Hong Kong
Printed by Imprimerie Aubin, Poitiers

ISBN 92-3-101922-8

© Unesco 1981
Printed in France

Preface

Over the past twelve years Unesco, in collaboration with the Scientific Committee on Oceanic Research (SCOR), has synthesized, through this series of 'Monographs on Oceanographic Methodology', the available field and laboratory research techniques that are needed to examine some of the most relevant marine scientific problems. Such syntheses are a vital component of the modern process of technology transfer.

The six preceding titles in the series have described techniques related to zooplankton sampling, zooplankton fixation and preservation, primary production, photosynthetic pigments, coral-reef research, and phytoplankton.

As the major marine advisory body to Unesco, the Scientific Committee on Oceanic Research (SCOR) prepares the manuscripts for the series. Most certainly, without this collaboration, dissemination of current research information to the scientific community would be impaired.

Following recommendations from SCOR that a given topic is of particular relevance to current research trends, international experts are appointed to a joint working group to decide which approach is best suited to the subject in hand and to prepare the detailed methodological descriptions. Thus, by the time a manuscript is submitted to Unesco for publication, the SCOR specialists have spent long hours of research, comparison and review as their personal contribution to international marine science.

Unesco is highly appreciative of the efforts of the scientists who prepared the present volume and wishes to express its particular thanks to Drs T. Platt, K. H. Mann and R. E. Ulanowicz, the editors, for their devotion to this project.

The scientific opinions expressed in this work are those of the authors and are not necessarily those of Unesco. Equipment and materials have been cited as examples of those most currently used by the authors, and their inclusion does not imply that they should be considered as preferable to others available at that time or developed since.

Contents

Introduction

This volume represents the first fruits of the labours of the SCOR Working Group 59, 'Mathematical Models in Biological Oceanography (with IABO)', which was set up in 1977. The terms of reference included the following:
To suggest mathematical methods in marine ecology for the design of research programmes in the open sea and the near-shore waters.
To suggest experiments for the treatment of biological-data collections with particular reference to the development of mathematical models.
The following served on the working group: K. H. Mann, Chairman; T. Platt, Vice-Chairman; J. M. Colebrook, M. J. Fasham, J. Field, V. V. Menshutkin, G. Radach, D. F. Smith, R. E. Ulanowicz, M. Vinogradov, F. Wulff.

The first meeting was held at the Institute of Oceanographic Sciences, Wormley (United Kingdom), 6–9 December 1977 and was attended by Mann, Platt, Colebrook, Fasham, Field, Smith, Ulanowicz and Wulff. J. W. Horwood was invited to be present to contribute expertise in fisheries modelling. It was decided to embark on a review of the state of the art in biological oceanographic modelling, with particular attention to new and promising developments in this youthful and rapidly advancing field.

Drafts of sections were prepared and brought for discussion to a second meeting, held at the Bellairs Research Institute (Barbados), 19–24 February 1979. Present at this meeting were: Mann, Platt, Colebrook, Fasham, Radach, Smith, Ulanowicz and Wulff. J. Field, who was prevented from attending this meeting, met with various working-group members soon afterwards, and made useful criticism of the draft manuscript. There is a sense in which the whole volume was a product of the working group's deliberations. The meetings were marked by a high degree of intellectual excitement, and the decision to include in the book a major section on holistic approaches, which is something of a novelty in this field, was reached unanimously and pursued with enthusiasm. Nevertheless, indications of the names of the persons responsible for writing drafts of each section and for responding to the criticisms of the working group are given as footnotes at the beginning of those sections.

At a third meeting, held in the Bedford Institute of Oceanography, Dartmouth (Canada), on 9–13 July 1979, Mann, Platt and Ulanowicz met to fill gaps in the manuscript and begin the editorial process.

ACKNOWLEDGEMENTS

The working group wishes to thank the following for assistance with the work: G. Hemmen, Assistant Secretary, SCOR, who worked untiringly to smooth all difficulties in our path; the Director of the Institute of Oceanographic Sciences, Wormley, who provided space and facilities for our first meeting; the Director, Bellairs Research Institute of McGill University (Barbados), who provided accommodation, meals, and excellent working facilities for our second meeting; J. Horwood who, after attending as a guest at our first meeting, provided written material which was most helpful and became incorporated in the writings of others; Marilyn Landry who typed the first and second drafts of the manuscript, sometimes working far into the night.

Finally, the committee wishes to thank the Executive of SCOR, who made possible some very fruitful scientific exchanges and stimulated several members of the working group to describe the meetings as the most interesting project they had been involved in.

Guide to the contents[1]

The working group recognizes two approaches to modelling biological oceano-graphic systems: the reductionist and the holistic. The first of these begins with distinct processes that are studied in isolation from the total system, in the classical manner of laboratory science. Section 1.2 deals with process models, distinguishing between empirical models and rational models, and showing that empirical models, while often useful for immediate practical application, need to be modified as early as possible in favour of models incorporating information on the mechanisms involved.

There are, however, many questions in biological oceanography that involve the interactions of several processes working simultaneously. These may be represented mathematically by coupled process models. Simulation of the complex set of interactions on a computer is a technique which was developed for modelling physical systems, but which is now widely used in biology. Section 1.3 deals with this approach, and Section 1.4 discusses some of the operations involved in applying systems-modelling techniques to eco-logical problems.

There is already an abundant literature on simulation modelling, but Chapter 2 of this volume breaks new ground in exploring some approaches to the holistic properties of ecosystems. Since ecosystems may be regarded as extremely complex networks of interactions between biological and physical elements, Section 2.1 deals with the possible applications of network analysis to ecological problems. Section 2.2 deals with another type of approach, namely through thermodynamics and statistical mechanics. While many of these ideas are in early stages of development, there is one theme which emerges from them all. It is:

For understanding biological oceanographic systems, it is necessary to have at least as much information on the fluxes as on the biomasses.

In the progress of biological oceanography through taxonomy and physi-ology to autecology and community ecology, there has, up to now, been much more information collected on biomass than on fluxes. This is easy to under-stand, since it constitutes a natural progression, but the time has come when

1. By K. H. Mann.

the balance must be redressed and large bodies of data collected on the fluxes of materials and energy through oceanographic systems. As Section 3.1.1 states, the most common fluxes are: (a) trophic transfers, in which production at one level is transferred by predation, grazing, egestion, detritus formation, etc.; and (b) cycles of elements which are mediated by plant-nutrient uptake, excretion, and such physical transports as nutrient upwelling, etc. Most fluxes are the product of a physiological rate and a population density, and Sections 3.1.2 to 3.1.4 deal with techniques for measuring fluxes.

Biological processes are profoundly affected by physical and chemical oceanographic phenomena, and one of the difficult problems to be faced is that of reconciling phenomena which occur at vastly different scales in space and time. Is it possible, for example, to include, in the same model, microbial events which occur on a scale of μm and seconds, and changes in physical oceanographic patterns which occur on a scale of, perhaps, 10^5 m and 10^9 seconds? These questions are raised and discussed in Sections 1.4.1, 3.2, and 3.4.1. In Section 3.3 the whole question of the way in which physical transports interact with biological processes is explored. Upwelling, stratification, ring and eddy formation, as well as the better-known horizontal ocean currents, all have strong, even overriding influences on biological events; but there is much work to be done before we have data on these phenomena which are in a form suitable for integration with biological data. The technology of automatic data acquisition in biology is still in its infancy; the need to pursue this line of work is discussed in Section 3.4.

Finally, the question of how to evaluate stress on marine systems is considered in the light of current ideas on ecosystem properties. It is shown, for example, in Section 3.5 that indices of diversity are probably well correlated with fundamental ecosystem properties, but in the ideal world a diversity index would include all organisms in the community, instead of just a small subset, as in most published studies. The merits of this and other approaches to ecosystem stress are tentatively evaluated.

1

State of the art

1.1 The classes of models in biological oceanography[1]

Mathematical models range in scope from simple representations of single processes, e.g. physiological, behavioural or demographic, through models of several interacting processes, to attempts to represent mathematically the behaviour of whole ecosystems. The process models are interesting and worth reviewing, but on the whole are well accepted by the oceanographic community, so that discussion is not so much about the usefulness of that class of models as about the relative merits of various mathematical approaches. By contrast, ecosystem models are controversial. There are many who work at the level of organisms or populations who doubt the value of ecosystem modelling and there are those who, having constructed a particular type of ecosystem model, have difficulty explaining to the first group in unambiguous terms just what scientific progress they have achieved. In this review of the state of the art we intend to deal with both ecosystem models and process models in the light of current thinking, and in particular we hope to show some of the implications of recent developments in ecosystem theory, for the future development of modelling.

ECOSYSTEM MODELS

The field of fish population dynamics is sophisticated and rigorous, yet there are many examples of the need to consider interactions several steps removed from the stock of interest. Cushing (1961) re-analysed the failure of the Plymouth herring fishery and found that recruitment had been adversely affected by a competing food chain, which led to depletion of dissolved phosphorus at the time of year when it was needed for the food chain leading to the herring. All the elements of an ecosystem were involved in this story: phytoplankton, zooplankton, fish larvae, nutrients, and by implication the physical factors responsible for the upwelling and advection of the nutrients. Similar examples could be drawn from fisheries in the upwelling areas of the world (Cushing,

1. By K. H. Mann.

1971) or from the Pacific sardine fishery (Murphy, 1966). Economically important changes in stocks of marine invertebrates can be shown to have their origins in multistage ecosystem effects. Declines in stocks of American lobster have been shown to be related to a four-step interaction along the food chain seaweed → sea urchins → lobsters → man (Mann, 1977), with ramifications to several other components of the system.

It is clear, therefore, that if a major decline of a fish stock is caused by events which take place in several parts of an ecosystem it will not be possible to understand, anticipate or modify such changes by confining attention to the population dynamics of the fishery. There are good economic reasons why attention should be focused on a higher level of organization, the ecosystem. Many other applied problems, such as eutrophication or heavy-metal pollution, could benefit greatly from this approach.

COMPARTMENTAL FLOW DIAGRAMS

As soon as we begin to try to quantify system interactions, we come up against the need for ecosystem models. At the simplest level, we may wish to make compartmental flow diagrams indicating qualitatively and perhaps quantitatively what interactions are occurring in the system of interest. To do this, it is necessary to represent all fluxes in the same units, such as carbon, or nitrogen or energy, so that plant biomass, food intake, faecal production, respiratory loss, etc., may be compared. Figure 1 and Table 1 (from Brylinsky, 1972) show how some quite old numerical data on the English Channel may be transformed into a compartmental flow diagram by the addition of some literature values on calorific equivalents and metabolic rates. From this we can see at a glance some features that were not apparent in the table. For example, phytoplankton production is about equally divided between zooplankton and benthos, but the benthos also receives a large contribution from the zooplankton. The output of zooplankton to other trophic levels is far greater than that of the benthos, indicating a greater ecological efficiency in the zooplankton.

In this diagram there is only one average value for each flux, so at this stage it is a description of some hypothetical steady state. If we wish to consider the mechanisms producing year-to-year variations, or the effect of perturbing the system, we need to go to a simulation model.

COMPLEX SIMULATION MODELS

Techniques for converting a compartmental flow diagram into a dynamic simulation have been borrowed from systems science and were used for modelling physical systems long before they were used in ecology. Basically, the technique involves writing equations to represent the transfers between the compartments. Having set the initial values of the state variables in the compartments, it is then possible to compute the effect of transfers into and out of the compartments as time progresses.

14

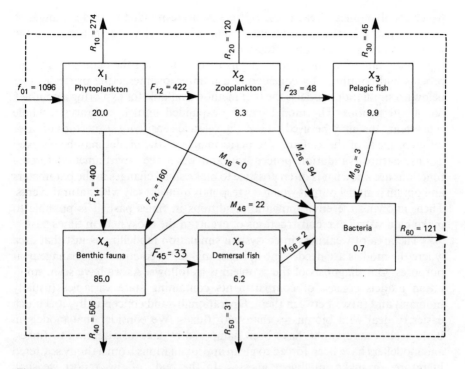

Figure 1
Block diagram of energy flow for the English Channel. Standing crops are in kilocalories per square metre and energy flows are in kilocalories per square metre per year (from Brylinsky, 1972)

TABLE 1. Summary of data[1] on the English Channel taken from Harvey (1950)

Organism group	Standing crop	Daily production	Daily respiration
Phytoplankton	4	0·4–0·5	—
Zooplankton	1·5	0·15	4
Pelagic fish	1·8	0·0016	1·25
Demersal fish	1·00–1·25	0·001	1·25
Benthic fauna	17	0·03	1·25–2·00
Bacteria	0·14	—	30

1. Standing crop and production values are in grams dry organic matter per square metre. Respiration is presented as percentage of standing crop.

In the example from the English Channel, transfer coefficients were listed for feeding, respiration, and mortality of each compartment, and the assumption was made that transfer of energy from a donor compartment to a receiving compartment was proportional to the amount of energy contained in the donor compartment. This is a very simple linear assumption and takes no account of variations in temperature, nutrient supply and so on. Nevertheless, it was possible to obtain a time series of each state variable and explore the way in

15

which the dynamics of each part of the system responded to small changes in all other parts. In the appropriate section of this review we shall explore the value of making models along these lines.

In the line of inquiry which begins by investigating the details of a single process and continues by considering a number of interacting processes (see below), a simulation model is the best method available for exploring a complex set of interactions. The model may be expanded until it embraces a whole ecosystem, and may be used to test the relative sensitivity to perturbation of different parts of the system. The performance of the model may be checked against nature (validation procedure), and when the environmental forcing functions are known it is often possible to make small changes in the parameters and obtain a model output which corresponds quite closely with natural events. Then, knowing the environmental conditions in times past, it is possible to make hindcasts, i.e. reconstructions of events in the ecosystem in times past.

The area of weakness in ecosystem simulation modelling is just that area where the models are needed most, i.e. in making predictions for management purposes. Our diagnosis of the problem is as follows. As we have seen, simulation models consist of compartments containing state variables (usually biomass) and fluxes between them. Operationally and conceptually, it is much easier to deal with biomasses than with fluxes. We consider that models in biological oceanography have been hampered by lack of attention to fluxes, and modellers have been forced to glean approximations from a thinly scattered literature, or make intelligent guesses. In the body of this report we shall explain in detail what we think should be done to remedy this lack.

Relatively simple versions of ecosystem simulation models have been both deterministic and spatially homogeneous. That is to say, the parameters of the fluxes are derived from constant average values for specified time periods such as a year, with no provision for variation about the mean. Moreover, the biomass values in the compartments were assumed to be spatially homogeneous, so that there was no provision for patchiness in the distribution of, for example, phytoplankton or zooplankton. As every biological oceanographer knows, constant relative rates of production or grazing and homogeneous distributions of organisms in the sea very seldom occur, and some of the more recent models have been cast in stochastic rather than deterministic form, and have made provision for spatial heterogeneity.

HOLISTIC APPROACHES TO ECOSYSTEMS

There are other problems in the use of compartmental simulation models, some of which are widely recognized. One problem concerns the definition of the state variables; how many compartments to have, and what to put in them. There is a conflict between the view that all species, indeed all individuals, have slightly different properties, so that it is desirable to put each species in a separate compartment, and the contrary view that at this degree of complexity the model is both unwieldy and lacking in general application. A second problem is concerned with time scales. The bacteria in Figure 1 may go through many thousands of generations during one generation of a fish. A small change in the properties of the bacterial compartment may have enormous and unrealistic consequences for the model as a whole.

Division of an ecosystem into compartments, and the attempt to re-create the properties of an ecosystem from its components, is a reductionist approach. The alternative is to consider from the beginning the properties of the total ecosystem. This is the holistic approach. Implicit in this is the view that ecosystems, like organisms, have a hierarchy of structures, and that at each level in the hierarchy there are properties that are characteristic of that level. The series of hierarchical levels that included molecules, cells, organs, organisms and populations is well understood and accepted, but there has been a reluctance to recognize that an ecosystem is a level of organization higher than that of populations, and that it has emergent properties.

In other disciplines the properties of systems are objects of study in their own right. In this review we shall consider the consequences of dealing with some of the unique properties of ecosystems. These include their thermodynamic or statistical properties, their size spectral composition, and various aspects of their food-web structure.

MODELS OF SINGLE PROCESSES

The case has been made that it has often proved desirable to have an understanding of total system function in a particular area of ocean. There are many other situations in which attention should be focused very precisely on one process in an ecosystem, and interactions with other parts of the system should be treated very generally, or in some cases ignored. Examples are studies of the relationship of photosynthesis, respiration, growth, production, feeding, or behaviour to one or more environmental variables. The information may be needed for some immediate application, or it may be a way of gaining insight into the workings of marine systems generally.

From the point of view of the laboratory biologist, the ideal situation is one in which all significant variables are observable and capable of experimental manipulation. An example would be the feeding rate of an animal, in which the relevant variables might be quality and quantity of food, temperature, light intensity and past history of feeding. Under laboratory conditions all these variables can be controlled, and plots can be made of the relationship between feeding rate and any one of them. Simple regression models may be used to express the relationships of feeding rate to each one in turn, or multiple regression models may be used to represent several variables simultaneously.

With laboratory data, the question frequently arises as to whether the results can be extrapolated to field conditions. It is possible that some variables have been overlooked or are not capable of being reproduced in the laboratory. In the example of the animal feeding rate, it is possible that the food is

17

distributed in nature in a complex manner which affects the accessibility to the animal, and hence the feeding rate. At this point it becomes advisable to collect data from the field if possible, but we then find that the environmental data can be observed but not controlled. The observer is driven to analysing whatever combinations of factors happen to occur naturally, and may find that there is a much less complete data set. The types of data analysis used are liable to be much the same as with laboratory observations, but the statistical significance of the conclusions may be at a much lower level.

MODELS WITH SEVERAL PROCESSES COUPLED

An example of a group of processes which interact might be those which influence secondary production in the benthos. Since the rate of secondary production for a population of animals is given by the integral of the product of number of animals and growth rate per animal, any process which influences either population density or growth rate will be of interest. The list might include feeding by the benthos, assimilation, growth, competition between organisms, reproduction, recruitment, and loss by predation. It may be possible to investigate the relationship between pairs of processes and express the results by means of appropriate regression equations, but when the whole complex of interactions is considered, a more powerful method of analysis is needed. One useful approach is to construct a general simulation model of fluxes, which gives time series for gains and losses in benthic biomass. The techniques are basically similar to those used for general simulations of ecosystems, and the two kinds of models grade into one another.

CHOOSING THE APPROPRIATE MODEL

Choice of model requires first of all that the objectives of the modelling be precisely defined. If the objective is an applied one concerned with the management of a stock or perhaps the management of an ecosystem to minimize the disruptive effects of human interference, then it is clear that some kind of prediction about the consequences of different management strategies is being sought. Under these circumstances a relatively simple model relating the process of interest to a few environmental variables may well be the best choice. If it appears that several interacting processes may be involved, the solution of the problem may require construction of a compartmental flow diagram of interactions and a computer simulation to explore the consequences.

If, on the other hand, the objective of the work is not tied to an immediately pressing management problem, but is part of the search for a better understanding of processes in biological oceanography, consideration should be given to modelling the whole ecosystem under investigation. The case is made, in the body of this review, that ecosystems have distinctive properties not found at lower levels in the hierarchy, and that there is a pressing need in biological oceanography for innovative approaches to the understanding of ecosystem properties. There is a wide range of approaches from which to choose. The one which up to now has received most attention is construction of compartmental flow diagrams and dynamic simulation of time series of state variables. However, review of achievements in this field suggests that there are enough difficulties with this technique to make it well worth considering the alternatives. Variants on the traditional approach to food-web structure should be explored, and the use of size spectral analysis as an alternative to taxonomic grouping offers exciting possibilities. Some solutions to the problems resulting from inhomogeneities in distributions of organisms in both time and space are being found through application of spectral analysis.

Each of the new approaches carries implications for the design of sampling programmes, and these too are reviewed in what follows. The long-range goal to which we are all working is the ability to make valid predictions about the working of marine ecosystems and about those attributes for which the systems are being managed. We are convinced that before this goal is reached, work on the holistic properties of ecosystems will have to be accepted as one of the major activities of biological oceanography.

1.2 Process models[1]

In this section we discuss some aspects of the modelling of particular processes. Implicit in the use of the word process is the conviction that the dynamics can be described through one or a few parameters. Often, however, use of a small number of parameters is merely a reflection of the chosen level of description: parameterization can be, but is not necessarily, a way of concealing our ignorance of how things work. The processes that we are talking about might be physiological, behavioural or demographic. Frequently, a number of these unit process models (submodels) are combined or coupled together to produce a model of a larger system, perhaps an entire ecosystem. Some examples of these unit processes are: the relationship between zooplankton grazing and the concentration of available food (Parsons et al., 1967); the photosynthesis of phytoplankton as a function of available light (Jassby and Platt, 1976); and the growth of fishes in terms of their ration and activity (Paloheimo and Dickie, 1966).

EMPIRICAL MODELS

Empirical models of a process are constructed with little or no reference to its internal mechanism (often unknown). Their function is descriptive: they are used to organize and summarize bodies of data relating two or a few variables. The empirical quantitative description may then be applied as a predictive tool. To the extent that internal mechanisms are not understood, either

---

1. By T. Platt.

in detail or through the identification of relevant macroscopic properties, finding a suitable empirical description is equivalent to the curve fitting of an arbitrary equation.

When empirical models are used for prediction, the precision of the estimates can be improved by increasing the number of arbitrary parameters to be fitted. Typically, however, these supernumerary parameters have no ecological interpretation. In the most extreme cases, data are fitted to generalized functions, such as polynomials of arbitrary order, and the data can be made to fit the 'model' exactly if the number of arbitrary constants (= the degree of the polynomial) is equal to the number of degrees of freedom in the data. But it will not be possible to identify these parameters with any known properties of the system under study: they are merely a set of numbers that give a good fit. Lack of structure in the model means that predictions outside the range of the independent variable for which we have prior data are quite unreliable. With any degree of fit the quality of the model is limited by the quality of the data: any model fitted to the data, e.g. by least-square methods, which yields model values within the error bars of the data points is of comparable quality, as long as no further knowledge about the process enters the model. The model description will only become essentially better when the data have smaller error bounds.

It is possible, however, even when working with models that are largely empirical, to choose a formulation in which the parameters have a definite biological meaning. This will be brought out in the discussion below on models relating photosynthesis by phytoplankton to available light.

RATIONAL MODELS

At the other extreme from purely empirical models are rational models: these are based on what is known or assumed to be known about the way in which the system is structured. With rational models, we expect the results to be more generalizable than with empirical models, and we might be less cautious about extrapolating outside our range of experience with the independent variable. The more we know, the less arbitrariness we can admit, until, if we know everything, we should find that writing down the implied model is a fairly trivial task.

SEMI-EMPIRICAL MODELS: A CASE STUDY

The more typical process model lies somewhere between the wholly empirical and the wholly rational. It exploits what is known about the structure of the system to minimize the arbitrariness in the model (the principle of parsimony), then uses empirical data to flesh out the description. The skill here is in choosing the parameters so that, once fitted, their numerical values tell us as much as possible about the biology of the system under study.

We illustrate the procedure involved in developing a semi-empirical

20

process model and testing it against alternative formulations by describing the evolution of models of the relationship between photosynthesis and light for phytoplankton.

The basic form of this relationship has been known for at least seventy years. If phytoplankton photosynthesis is increased at a series of light intensities, the plotted data look something like Figure 2(a); an increase at low light intensities up to some plateau value. Sometimes the data do not appear to pass through the origin, but intersect the ordinate at some negative value (Fig. 2(c)). If the light intensity is increased to values approaching full sunlight, it is found that photosynthesis begins to decrease (Fig. 2(e)).

Curves of general shape shown in Figure 2(a) are called hyperbolic. The simplest approximation to this curve, and the earliest (Blackman, 1905), consists of two straight lines: one describing the plateau, the other describing a linear increase up to this level (Fig. 3). Most of the commonly used curvilinear forms are discussed in Platt *et al.* (1977) and in Jassby and Platt (1976).

Mathematical description of hyperbolic equations requires a minimum of two parameters, or three if the curve does not pass through the origin (e.g. Fig. 2(c)). Selection of the parameters to use is a critical step in the development of a semi-empirical model; we rely on our intuition and on what is known about the mechanics of the process under study. There is no consistency at all in the selection of parameters by workers who have constructed models of the light-saturation curve.

Ross (1970) lists the criteria of well-chosen parameters as (a) small variance; (b) lack of correlation with other parameters; (c) interpretability. It is known that photosynthesis is a multistage process, involving light-independent reactions as well as light-dependent ones. The quasi-linear part of the curve at low light intensities may be parameterized by its slope α, which is related to the quantum efficiency of photosynthesis (Fig. 2(b)). To the extent that α characterizes a photochemical reaction, it should be independent of temperature. The height of the plateau P_m (Fig. 2(b)), on the other hand, is thought to depend on the dark reactions of photosynthesis and to be controlled by rate-limiting enzyme reactions. We might therefore expect it to be temperature-dependent. The intercept on the ordinate, called R, is interpreted as the dark respiration of phytoplankton.

These three parameters, α, P_m, R, may serve as one parameter set to describe the curve relating photosynthesis and light (Platt *et al.*, 1977; Jassby and Platt, 1976). They are by no means the only possible choices. Figure 4 shows some of the alternatives. The parameter I_k is the value of I for which the extrapolation of the linear portion intersects the plateau; I_s is the value of I for which P is half its maximum value; I_m is the value of I at which P is a maximum.

Given that different authors parameterize their light-saturation equations in different ways, a preliminary step in the comparison of equations is to recast them so that all are consistent with respect to the parameters used (Platt *et al.*, 1977; Jassby and Platt, 1976). If we adopt α, P_m and R as parameters,

21

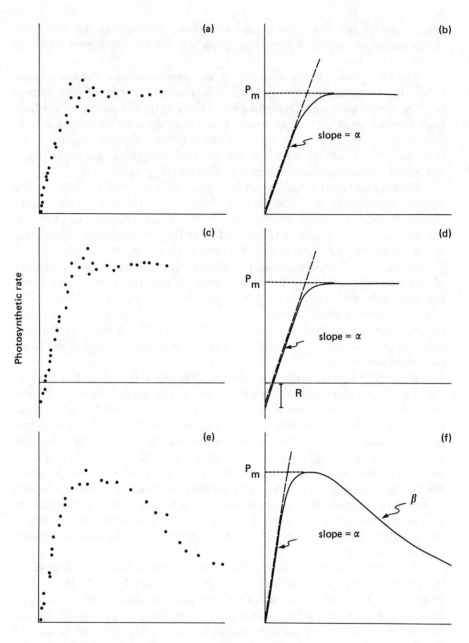

Figure 2
Typical data (a), (c) and (e) and corresponding theoretical representations (b), (d) and (f) of the relationship between photosynthesis and light for phytoplankton. Panels (a) and (b) show light saturation; panels (c) and (d) include an intercept for respiration, R; panels (e) and (f) include photoinhibition

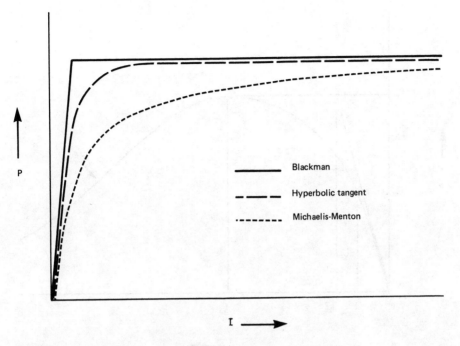

P

I ⟶

Figure 3
Three well-known descriptions of the light saturation curve

the procedure is: (a) differentiate the given equation with respect to I; (b) calculate $\partial P/\partial I$ for $I \to 0$, and set this equal to α; (c) set $\partial P/\partial I = 0$ and find the maximum P_m; and (d) R will be the value of P at $I = 0$.

These steps will be sufficient to eliminate other parameters from the given equation, unless the equation has been overspecified (too many arbitrary parameters). Alternative representations of the light-saturation curve can thus be put into common form for comparison.

Before the models are applied to data, one further step is necessary: to normalize the data so that the effect of differences in biomass between samples (stations, times of year, depth) is removed. Taking chlorophyll a concentration as the appropriate biomass index (Platt *et al.*, 1977), we normalize by dividing the production rates by chlorophyll, to give normalized production rates P^B. With this normalization, the new parameter set is α, P_m^B and R^B.

Because of the limited structure of these semi-empirical models, their sole criterion of merit has to be their ability to fit experimental data. Jassby and Platt (1976) compared eight three-parameter formulations of the light-saturation curve using data from 188 experiments on natural assemblages of coastal phytoplankton. Two indices of goodness-of-fit were calculated: first, the mean squared deviation between measured and calculated production for each equation, averaged over all experiments; second, the mean-squared deviation for each experiment, scored over all experiments. Both criteria led

23

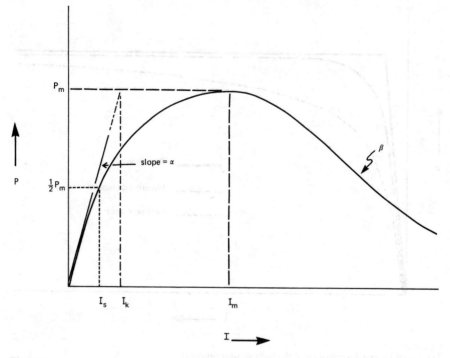

Figure 4
Various parameters used in theoretical representations of the light-saturation curve

to the same conclusions. The equation,

$$P^B = P_m^B \tanh(\alpha I/P_m^B) - R^B, \tag{1}$$

was found to give the best fit most of the time (Fig. 3). Conversely, a Michaelis-Menton representation, commonly used in coupled process models, was found to give a relatively poor fit to field data (Fig. 3).

Concentration of available nutrients is another possible source of variance in measurements of P^B. The effect of nutrients may be introduced into the predictive equation either through their effect on the light-saturation parameters (Platt *et al.*, 1977) or directly, as in Bannister (1979).

So far in this discussion, we have not treated the shape of the curve for light intensities higher than the onset of photoinhibition. The equation most commonly used in coupled process models when inhibition is to be taken into account is Steele's equation,

$$P^B = \alpha I_{\exp}\left(-\frac{\alpha I}{P_m^B}\right). \tag{2}$$

The weakness of this equation is that it contains no additional parameter to characterize the inhibition process. The slope of the inhibition portion of the curve is not independent of the initial slope α in this formulation. Although the

24

nature of photoinhibition is not understood, there is no reason to suppose that it is characterized by the quantum efficiency of photosynthesis.

If it is legitimate to characterize photoinhibition by a single parameter β (Fig. 2(f)), a suitable equation for describing the complete light-saturation curve is (Platt et al., 1980),

$$P^B = P_s^B(1 - e^{-\frac{I\alpha}{P_s^B}})e^{-\frac{I\beta}{P_s^B}} \tag{3}$$

where P_s^B is a scale factor related to P_m^B by the identity

$$P_s^B \equiv P_m^B\left(\frac{\alpha + \beta}{\alpha}\right)\left(\frac{\alpha + \beta}{\beta}\right)^{\beta/\alpha} \tag{4}$$

The advantages of not writing P_m^B explicitly into the main equation are first that the equation is then less complicated, and second that the partial derivatives of P with respect to the parameters, needed in non-linear curve fitting routines, are much simpler to evaluate.

CONCLUDING REMARKS

Our ultimate goal in process modelling is the construction of robust, rational descriptions. In practice, ignorance forces us to accept, at best, a semi-empirical description. It is expedient to abandon as early as possible the wholly empirical description in terms of generalized functions. Plotting of simple scatter diagrams of data on the variables of interest will suggest working hypotheses for the development of semi-empirical descriptions. The simplest such hypothesis is the assumption that the process is linear in the independent variable. For example, in modelling zooplankton grazing, we might as a starting-point make the assumption that ration increases linearly with available food. Experience in working with this model as a description of experimental data will show that this linear assumption is not robust at higher food concentrations such that the concept of a saturation level or feeding plateau has to be introduced (Ivlev, 1945). And then, detailed investigation of the feeding at very low food concentrations reveals the existence of a threshold food concentration below which feeding ceases. Such a threshold was introduced into Ivlev's model by Parsons et al. (1967), giving a model with three parameters: the magnitude of the threshold; the height of the plateau; and the speed with which feeding rate approaches saturation. Recent research has shown that the heights of the feeding plateau itself can vary in response to the available food (Conover, 1978; Mayzaud and Poulet, 1978). It should be possible to introduce this new information into the models through the idea of adaptation of the parameters, with a certain time constant, to changing environmental conditions.

The strategy of building semi-empirical models, then, is to start from simple assumptions suggested by the data; to incorporate as much as possible of what is known about the process; and to be prepared to modify the model in the simplest way that will be compatible with new information that becomes available.

1.3 Coupled process models

1.3.1 DIGITAL SIMULATIONS[1]

Introduction

As successful, rational and semi-empirical models of biological processes accrue, it is only natural that we should seek to link up the models in the same way that the prototype processes are related within the biological-physical community. The hope is that the system of coupled models will behave, at least in a qualitative sense, like the real ecosystem. Such aggregation is, after all, a fundamental tenet of reductionist systems theory—that the behaviour of the entire community can be fully described in terms of causal forces observable at the scale of the subsystem.

Thus, the past decade has witnessed a succession of simulations of freshwater (Park *et al.*, 1974; Patten *et al.*, 1975), estuarine (Chen, 1970; DiToro *et al.*, 1977; Kremer and Nixon, 1978; Radford, 1978) and marine (Jansson, 1972; O'Brien and Wroblewski, 1973; Nihoul, 1976; Steele and Mullin, 1977; Vinogradov and Menshutkin, 1977) ecosystems. While the various models differ in numerous details, they practically all include the interacting processes of nutrient uptake, production, grazing, predation, excretion, respiration, benthic exchange, advection and dispersion.

Perhaps the most complex model to be reasonably validated is that of DiToro *et al.* (1977) of the Sacramento–San Joaquin river system. The model simulates algal growth mediated by light levels, bacterial remineralization, zooplankton grazing and the physical effects of advection, dispersion and nutrient loading. Budgets of phosphorous, nitrogen, silica, oxygen and living biomass are maintained simultaneously.

Problems in setting up a coupled process model

There is no unique way to find out the characteristic properties of a system. The most important source remains intuition. Therefore, setting up a model is basically an arbitrary undertaking (Stahl, 1967). But there are methods which can support intuition.

Setting up a coupled process simulation model for a given problem goes through several steps (Nihoul, 1975):
1. Spatial and temporal limits have to be defined. We have to determine the geographical region and the time which we want to consider. This assumes the knowledge of the scales which govern the problem.
2. The complexity of the problem has to be determined by considering how the subsystem under study is embedded in the total system. State variables and areas in parameter space have to be defined.

1. By R. E. Ulanowicz and G. Radach.

26

3. The complexity of the problem has to be reduced by carrying only a very limited number of state variables in the model. Very often the state variables summarize quantities like *the* phytoplankton, etc.
4. Spatial and temporal variability have to be eliminated as far as possible, e.g. by averaging procedures (over depth, time or area) without affecting the scales of interest.

Some techniques that can help on the way from the natural system to the model system are interaction diagrams, statistical correlation analysis and scale analysis.

As can be seen from items 1 to 3 an essential problem of modelling an ecosystem stems from the fact that the ecosystem is an open system. It is not a closed unit. This is especially important for marine ecosystems, because the water movements transport substances and organisms from one habitat to another. We can only detect structural and functional focal points in the biosphere (Schwoerbel, 1977). This is one of the main reasons why the representation of *the essential* is a most subjective choice out of all available information.

This fact has important consequences for the construction of coupled process models (as for all complex biological models): the terms chosen to describe the ecosystem, the groups of organisms which can be combined into one state variable, and the degree of complexity which has to be incorporated are all subject to a personal bias.

A second major problem lies in the degree of complexity of the coupled process model which is *necessary* to treat the problem. Complexity is used here with no special formal definition in mind and may refer to the number of state variables incorporated, the number of single processes involved and the kind of representations for the interrelationships. There are no absolute criteria to determine that degree of complexity which is adequate for the problem and which enables us to reject a model as too complex or as too simple. The only possibility lies in defining the necessary complexity in relation to the data which form the basis for the posed problem.

A third big problem lies in the closure of the coupled process model. The restriction to a subsystem is, from the technical point of view, necessary to keep the problem manageable. The limits of the subsystem have to be defined in such a way that it is ensured that the connecting links—the interfaces—to the total system can be regarded as known. Thereby the network of the subsystem is closed. Often such loose ends at the 'border' of the subsystem are taken up by prescription of so-called 'forcing functions'.

The closure of a subsystem is achieved not only by regarding a concentration or a flux which connects the subsystem with the total system as known, but also by the formulation of all the other fluxes involved: all the formulations of the processes entering these coupled process models are presented in a way that only very few state variables are used to describe a complex process, e.g. nutrient uptake (Dugdale, 1977). The whole complexity of a single process is cast into a simple equation, and the constants in that equation are thought to represent all the important properties of the process bearing on the simulated ecosystem. Thus, because most of the details of the process itself have to be neglected, all these process models can be regarded as mechanisms to close the subsystem against the total ecosystem. This type of closure, referring to a special process formulation, is often called 'parameterization'.

Another problem enters the construction of coupled process models with the necessity to use conversion factors. This is a logical consequence of formulating a balance of energy or matter. Nevertheless, measurements obtained for comparisons with model results are given in different units. In every single equation, however, the units must be the same. This means that either all quantities are measured in the same units or conversion factors have to be applied.

Of course, conversion factors will merely create a new problem, if we cannot assume that they are constant over the time range considered in the simulation. For instance, it was shown that the spring phytoplankton bloom during the Fladen Ground Experiment, FLEX 1976, cannot be consistently modelled using a constant carbon-to-phosphorus ratio for nutrient uptake over the whole range of the bloom (Radach, 1979). If the exponential phase is modelled well, the post-bloom simulation fails. Similar problems seem to appear in other work (Sjöberg and Wilmot, 1977; Jamart et al., 1978; Kremer and Nixon, 1978).

There exists a great uncertainty in our present knowledge of the carbon-to-chlorophyll ratio and similar ratios (Banse, 1977). The reason seems to lie in the fact that data from different dynamic response patterns are used for the same correlation analysis. Essentially this critique comes down to the finding that a non-linear system cannot be described by linear regression analysis. What we need then are well-proven representations of these ratios in terms of other dynamic quantities, i.e. functional representations depending on the status of the system instead of constant ratios. So far at least, such well-proven parameterizations of conversion 'factors' in terms of one or a few state variables have not appeared. This problem shows again the difficulties involved in closing the system to be modelled.

Even if all prognostic variables in the coupled process model are formulated in the same units, it may well happen that the effect of a different state variable has to be introduced as a forcing quantity. Very often this problem is settled by formulating the influence of this quantity, e.g. nutrient limitation of phytoplankton growth, in non-dimensional expressions, where the units of the 'external' state variable (in this example: the limiting nutrient) cancel each other. Then no explicit conversion factor will appear.

Recent examples

Many interesting papers on particular coupled process models exist. We may be allowed to draw the reader's attention to a small selection: for upwelling ecosystems the papers by O'Brien and Wroblewski (1973) and Walsh (1975, 1977), for annual cycles the papers by DiToro et al. (1971, 1977) and Kremer and Nixon (1978), and for shorter-term events like plankton blooms the papers by Iversen *et al.* (1974), Winter *et al.* (1975), Jamart *et al.* (1978) and Radach (1979). It is nearly impossible to give a complete review.

There are several collections of modelling papers or reviews of the state of the art which have appeared within the last five years. Without any claim of completeness, we would like to mention here the series edited by Patten (1971, 1972, 1975); the two books originating from a conference on marine modelling, namely Nihoul (1975) and Goldberg *et al.* (1977); and the recent monograph by Kremer and Nixon (1978). These volumes compile so much material about models of special marine biological aspects that we can restrict ourselves in this section to brief comments on a few coupled process models to clarify some of the more general statements of the foregoing sections.

There are several models which could be chosen for this discussion. We concentrate first on the work by Kremer and Nixon because it is published as a book and easily available. Moreover, the model is sufficiently general to use it as an exploratory model for hindcasting and sensitivity analysis, and eventually for management purposes. Kremer and Nixon designed a model for synthesizing a great deal of information on Narragansett Bay, a coastal marine ecosystem. The numerical model simulates six state variables: phytoplankton, herbivorous zooplankton, ammonium, nitrate and nitrite, phosphate and silicate. The system is closed by the prescription of empirical relations between simulated and non-simulated trophic levels (bacteria, carnivorous zooplankton, fish, benthos).

The first aim they pursue is a simulation of the spatial variability of the dynamics of the planktonic system in the bay, by prescribing, e.g., annual cycles of light intensity, temperature and currents as forcing functions. Over fifty physiological constants, about ten annual cycles, and initial conditions determine the results of the simulation. The simulated horizontal gradients are of the same magnitude as in the data. But the time series of nitrogen compounds and phosphate are not sufficiently well reproduced. The authors then have used the model for 'an analysis of the bay ecosystem, to explore a number of hypotheses about which processes may or may not be important in forming the pattern that emerges in the natural system', for example, the assumption of a time-varying carbon-to-nitrogen ratio or the inclusion of dissolved organic nitrogen. Although no final answer was offered, directions of continuing research were shown.

This example evidently demonstrates the usefulness of complex simulation models: they can serve as more objective tools for synthesizing information and for revealing gaps in the current knowledge than can mere imagination. Moreover, this discrepancy between measured and simulated nutrients emphasizes the problem of 'sufficient complexity' of the model. It seems that dissolved organic nitrogen compounds probably have to be added to the prognostic state variables to enable a better agreement to be made between model results and measurements. It is the lack of data that prevents an immediate test of this hypothesis.

29

Another example of the usefulness of such models arises from the result that there is not sufficient phytoplankton available to feed the zooplankton. The same result occurs in the simulation of the spring phytoplankton bloom during the Fladen Ground Experiment, FLEX 1976, in the Northern North Sea (Radach, 1979). We might conclude that the grazing demand is over-estimated by the process formulation or that other food sources must be introduced. This leads immediately to two directions of research, namely (a) calculating the grazing demand not from a global quantity 'zooplankton', but from species and age stages with specific grazing rates, or (b) modelling of a herbivorous zooplankton system like Steele and Mullin (1977) propose, to parameterize the grazing pressure in a new way. In both cases, well-planned field experiments to determine the abundance at species and age stage level, well resolved in time over a few months, are indispensable.

Coupled process models, validated or not, can be used to detect the important links between components of the system or determine constants which have to be measured extremely well. Kremer and Nixon (1978) showed, for instance, that the annual temperature cycle is basically responsible for the dynamics in the Narragansett Bay model. If they replace the observed annual cycle by a mean temperature, the phytoplankton characteristics are partly lost and the cycle in total zooplankton is destroyed. This says at least that the model formulation is very strongly coupled to temperature, and that temperature is one of the most important forcing functions.

A further example concerns vertical turbulent mixing. Riley et al. (1949), Sverdrup (1953), Steele and Menzel (1962), Radach and Maier-Reimer (1975) have shown the importance of the turbulent vertical mixing process for the development of the spring phytoplankton bloom. It appears that there exists an optimum degree of turbulent mixing causing an optimum bloom. For smaller turbulent mixing, nutrient depletion and light inhibition quickly terminate the bloom. Too much vertical mixing causes too short residence times of phytoplankton and thus an optimum development is inhibited. These and other results were responsible for the intense physical programme in the frame of the Fladen Ground Experiment, FLEX 1976, in the North Sea to obtain physical data for the calculation of local turbulent diffusion coefficients as a function of time for use in the FLEX biological models.

The model by Kremer and Nixon will also be used for management decisions. In this case one has always to keep in mind the authors' following remark: 'If the natural system is changed in some fundamental way, the conceptual model that lies behind all of the equations and computer programming may no longer apply.'

Although these kinds of models may yield conceptually realistic output, we cannot fail to notice that the different submodels of such a complex model may be in a very different state of validation. Physical simulation models for tides and for the dispersion of a passive pollutant or heat are far more advanced than the biological submodels, and predict future states with a high degree of reliability.

30

Such a physical model forms the basis of another example of a whole-system modelling endeavour involving a significant commitment of management, manpower and resources: the general ecosystem model of the Bristol Channel and Severn Estuary (GEMBASE) being created by the Institute for Marine Environmental Research (IMER) at Plymouth (Radford and Joint, 1980). It follows the commonly accepted strategy for developing large systems simulations in that work usually proceeds in four stages.

The important first step is the agreement by all involved on the identity of the major elements of the system to be simulated. This step is often neither obvious nor trivial (see Halfon, 1979). Since modelling each individual organism, or even each taxon, is impractical, an appropriate level of aggregation needs to be chosen. In GEMBASE aggregation was done on the basis of both trophic level and habitat, resulting in seventeen state variables (see Fig. 5). The contents of each compartment are measured in terms of some common medium (carbon or nitrogen in the present version).

Deciding upon the significant exchanges in medium or energy between the various compartments forms the second set of major decisions to be made. Exchanges are indicated in the figure by the arrowheads. At this point one has created a conceptual model (see Section 1.3.2) of the system.

Conceptual models can be created without regard to spatial heterogeneity, but it often happens that different regions of the physical domain of the ecosystem differ markedly in biological behaviour. This is certainly true of the estuarine system being modelled in GEMBASE. In such cases, the domain is broken up into subregions (seven in this example); and, in principle, separate models might exist for each subregion. Exchanges occurring between subregions as a result of water movement must be included in the model. GEMBASE is, therefore, supported by two hydrodynamic models which are run independently on shorter time and distance scales, with the integrated exchanges being fed as input to the slower-scaled ecological models.

The crux of the modelling process is the mathematical description of the intercompartmental flows. Each flow may be the result of one or more processes, and each process must be described as in the case study in Section 1.2. For example, the flow of nitrogenous nutrients to the phytoplankton incorporates descriptions of the functional dependencies upon light, nutrients and temperature. In turn, each process may be characterized by more than one parameter. GEMBASE currently treats 150 processes involving some 225 parameters. As a consequence, a modelling exercise of this extent relies heavily on the 90 per cent of the parameter values which have to be taken from the literature. The limitations of such an endeavour are set by the data available to validate the model, usually not by mathematical and computational restrictions.

Upon completion of the process descriptions the model consists of a set of coupled, first-order differential equations, which must be integrated to yield time series for the separate state variables. There are several options here involving analogue or digital techniques. Practically all large-scale models

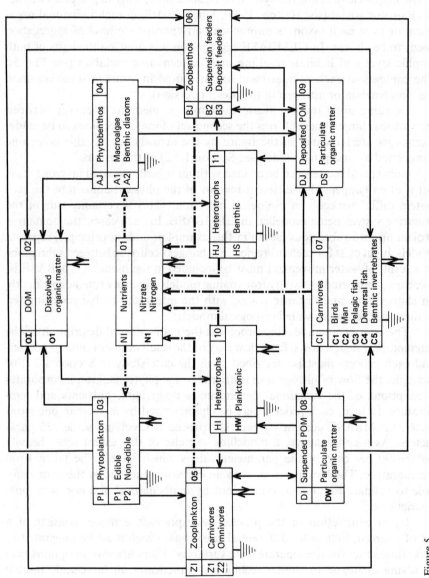

Figure 5
Process flow diagram for GEMBASE with water column to left, substrate to right (from Radford and Joint, 1980)

are now treated by digital algorithms. A number of higher-order programmes exists for simulating such mathematical systems, including CSMP, DYNAMO and MIMIC—all of which require very little mathematical expertise on the part of the modeller. The GEMBASE team has chosen to simulate in CSMP.

It helps greatly if the variables, flows and processes are indexed by some systematic scheme, an excellent example of which is the one introduced by the GEMBASE group. It allows the entire model to be presented verbally and mathematically as a series of process descriptions in the form of a manual.

The final stage of the project involves comparing the simulation output with data taken from the prototype. At this point the model is 'tuned' by varying parameter values until reasonable agreement is achieved between model output and data. The tuned model is then ready for testing (see Section 1.4.3) against a set of data independent of that used to create the model. Failure to reasonably validate the model (see Section 1.4.3) is cause to return to one of the previous steps and amend the model. Iteration should continue until verification is achieved on a completely fresh set of data. GEMBASE has now been calibrated for a three-year period. The ease with which the model was calibrated was most encouraging. IMER is now acquiring an independent three-year series of data for the purpose of testing the derived model.

General considerations

Whole ecosystem models are usually created with two objectives in mind: (a) to organize data-acquisition efforts from entire ecosystems and thereby discover hitherto unknown aspects about system behaviour, and (b) to be able to predict the behaviour of the ecosystem under arbitrary environmental conditions—ultimately in a quantitative fashion.

The first goal is subjective in nature, and there is probably little doubt that those undertaking the community modelling effort do benefit from the organization that any logical construct imposes upon their data-gathering effort. Impressive banks of synoptic data have been, or are being, gathered under the cited projects. Likewise, investigators are exposed to information on biological processes quite removed from their individual specialities. Thus, the educational value for those partaking in the investigation is undoubtedly great. In short, systems modelling has provided a convenient focus around which to conduct large-scale quantitative investigations of whole ecosystems.

However, the larger needs of science and society are not so easily satisfied by subjective accomplishments. Ultimately, ecosystems models will be held to the more objective criterion of predicting ecosystems behaviour. Most funding of systems modelling is done with this stated purpose. In fact, managing agencies are even beginning to demand systems modelling as part of environmental-impact assessment procedures. As much as we might wish to avoid the issue, we must eventually face the question of how well our models predict actual events beyond the scope of the data used to create them (see also Section 1.4.3).

It is in trying to answer this question that we are met with a seeming paradox. Those systems models with some degree of validation are biologically quite simplistic (Longhurst, 1978). It seems remarkable that Streeter-Phelps type calculations for BOD and DO should achieve a reasonable degree of predictive success for the Clyde (Mackay, 1973), the St John River (Sigvaldason et al., 1972), and numerous other riverine systems, given that the various biological processes establishing DO levels are treated in a relatively simplistic manner.

As one adds more processes and detail, the ideal of validation appears more difficult to achieve. That is not to say that acquired data cannot be simulated by highly complex models. On the contrary, by including a larger number of parameters into the system description, the number of degrees of freedom available to 'tune' the model to fit the data increases. Most of the above-cited expositions of models contain examples of good fits to real data. The results, however, may be quite sensitive (see Section 1.4.2) to changes in parameter values or initial conditions. Model behaviour outside the range delimited by the tuning data is frequently problematical.

Whence the question, why do some of the simpler coupled process models approach validation, while most detailed and complex representations appear to fall short? Although no definitive answer is available, some suggestions have been made.

Concerning the limited success of some simpler models, Mortimer (1975) attributes prediction ability to the circumstance that the organisms being modelled are 'biological actors on a physical stage'. This is another way of saying that the major controls upon a given biological variable are either physical (e.g. advection, light) or chemical (nutrients). Since the physico-chemical environment is more amenable to prediction with existing techniques, biological processes that are strongly forced by abiotic events can be reasonably predicted. When the major controls are biological in nature, however, problems arise in coupling biotic processes together.

One obvious difficulty with coupled process models appears at the outset of the endeavour. One must first choose the particular variables to be included in the model. The choice of what compartments to include is usually done according to some objective evidence available prior to the modelling endeavour. The problem often arises, however, where species change at 'a relatively rapid rate, leaving the ecosystem whose state was to be predicted a different one than that being modelled' (Dugdale, 1975). Such circumstance argues in favour of analyses which are relatively indifferent to the taxa present in the community, in much the same manner that thermodynamic theory is not predicated upon the existence of specific kinds of atoms and molecules.

Finally, it has been suggested (May and Oster, 1976; Ulanowicz, 1979a) that fundamental mathematical difficulties arise when treating coupled non-linear processes. Difficulties can be greater than a mere inability to achieve

34

an analytical solution to the model. Under particular parametric conditions some non-linear systems (Lorenz, 1963; Ulam, 1963) exhibit behaviour which is indistinguishable from chaos. That is to say, the system will remain within given bounds but never approach periodicity. Furthermore, the slightest change in initial conditions will result in a solution which, after a specified time, bears no coherence to the original solution. Since one is never able to measure the natural community with absolute precision, prediction in such systems is short-term at best. Most troubling is the observation (Tuck and Menzel, 1972) that increasing the number of coupled non-linear processes only increases the likelihood of incipient chaotic-like behaviour. The possibility thus arises that by including more processes in greater detail (this usually means *non-linear* detail) prediction ability wanes instead of waxes.

Attitudes towards coupled process models, therefore, appear polarized on the need for greater detail in models. Many feel that the modelling effort is still quite young, and that further experience and more precise, more complete data will eventually allow for reasonable prediction of community behaviour (e.g. Vinogradov and Menshutkin, 1977). Hardly anyone could disagree with the need for better data and data-collection strategies. Hence, Chapter 3 is intended to help all investigators plan data-acquisition programmes which will be of use, not only to the investigators' particular purpose, but to all those interested in describing processes in marine and estuarine systems.

None the less, a growing number of ecosystems investigators are beginning to question the wisdom of creating progressively more complex models. Others question whether the basic reductionist systems postulate is entirely applicable to whole ecosystems (Mann, 1975). Behaviour and taxa, after all, change with changing external conditions. The possibility remains that mechanisms and forces at the level of the subsystem change according to some holistic, or community, criterion. That is, they are 'selected for' over the course of time to satisfy some community principle in much the same manner that species are selected for in evolution. Lest such inclinations remain too abstract, we have chosen to explore selected approaches to holistic treatments of ecosystems in Chapter 2 with the hope that they will stimulate further efforts at articulating macroscopic ecological hypotheses.

1.3.2 CONCEPTUAL FLOW DIAGRAMS [1]

Although there is some disagreement about the value of the results of whole systems simulation, there is general agreement that the creation of conceptual models of interactions in ecosystems is a most beneficial exercise. By conceptual model is meant the translation of a 'word model' into a 'picture model' that summarizes present knowledge of the system in relation to the problem we intend to study. The conceptual diagram can sometimes be the final step in the modelling process, but usually the ambition is to carry the analysis further. Field

1. By F. Wulff.

and laboratory data are used to put numbers on storages and flows of matter of energy and, ultimately, a mathematical model is constructed for simulation on the computer in order to study the dynamic behaviour of the system.

The picture model must describe the state variables and the interactions within the system as well as the forcing functions. This can be done with a different degree of accuracy, depending on the purpose of the model and on the type of symbols used in the diagram. Simple 'box and arrow' diagrams may serve the purpose but often the ecologist, like the electronic engineer, needs a more specific set of symbols in order to describe his system. The circuit language for energy and matter developed by H. T. Odum is an example of such a sophisticated language, especially suited for ecological purposes. The development of these symbols follows from earlier attempts to summarize the information about whole ecosystems in conceptual diagrams (Odum, 1957). Fundamental to the use of the various symbols in the circuit language is the conservation of energy and matter, the degradation of energy into waste heat during work processes and the importance of feedbacks and interactions in the functioning of complex systems. A short description of the symbols is given in Figure 6. A complete description and the physical basis for the operation of the different symbols is found in Odum (1972). Many examples of how these symbols are used at different levels in the modelling process are found in the book by Hall and Day (1977).

Attempts to synthesize the information about a particular ecosystem may result in very complicated diagrams such as the Odum circuit model shown in Figure 7. The complex 'bird nests' or 'spaghettis' have been very much criticized (i.e. Hedgpeth, 1977), nevertheless they represent an attempt to look at the *whole* ecosystem instead of separate parts with no connections, the traditional approach taken by scientists from different disciplines. These models are not meant for computer simulations; they serve as 'book-keeping diagrams' in the planning of a multidisciplinary research programme where storages and flows are related to actual tasks in the field and laboratory. Conceptual diagrams are also important as a means of communication between scientists responsible for the data collection and those working with the computer simulations.

A conceptual diagram with numbers on storages and flows may pin-point the quality of the existing data on a particular system. The picture model of the benthic community in the Northern Baltic, made by Ankar (1977), may serve as one example. The flows of energy between the compartments in this model, shown in Figure 8, are derived from direct measurement of the macro- and meiofauna biomasses, multiplied by production/biomass ratios obtained from literature data. The attempt to describe the energy flows through the system gives a better understanding of the relative importance of different organisms than the quantitative biomass values alone. The calculation of energy flows through the food-web is an accounting process where sources and sinks of energy must be balanced and reasonable in terms of ecological efficiencies.

Another example is the description of the Black Sea pelagic system presented by Petipa *et al.* (1970). Field and literature data were summarized and showed highly variable ecological efficiencies, but many of the basic characteristics of the system were illustrated, without recourse to any sophisticated mathematics.

The pictures do not, however, explain why the ecosystem is organized in the way it is described. The importances of the different energy pathways may, at least in temperate waters, vary greatly during the year with successions of algal, zooplankton, benthic and fish species. Wyatt (1976) has criticized the interpretation of this approach. He considers that in the course of the life cycle of an organism it is part of several simple food chains which succeed one another. Construction of a single complex web obscures the succession and true mode of functioning.

1.4 Working with models

1.4.1 SCALING[1]

A general account of scale analysis is given in Section 3.2. Here we make just a few remarks relevant to scaling in dynamic simulation models.

When reduced to non-dimensional form, the simulation equation will be controlled by one or a few dimensionless groups of parameters. The entire range of behaviour of the system (as seen through the eyes of the model) can be demonstrated merely by varying these dimensionless groups through the expected range of their magnitudes for a substantial saving in computing time. In this way also, it can be seen clearly which of the terms of the simulation equation will dominate the equation under given conditions. An economical way to summarize the dynamics will then be to plot the simulated data for various values of the dominant dimensionless group.

The other main point to be kept in mind is that all processes on a finer scale than the grid scale of the model will necessarily have to be parameterized.

1.4.2 SENSITIVITY ANALYSIS[1]

Under the supposition that a particular model is a reasonable representation of the dynamics of the system concerned, we can ask the question, 'How sensitive (or conversely, robust) is the model to changes in the values of its rate parameters?' At a higher level of abstraction one can ask, 'How sensitive is the model to changes in the functional form of any or all of its component functions?' These are the problems of sensitivity analysis.

1. By T. Platt and G. Radach.

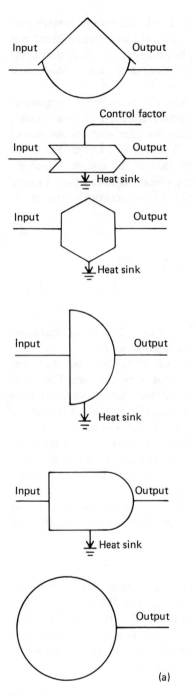

PASSIVE STORAGE symbol showing location in a system for passive storage such as moving potatoes into a grocery store or fuel into a tank. No new potential energy is generated and some work must be done in the process of moving the potential energy in and out of the storage by some other unit.

WORKGATE module at which a flow of energy (control factor) makes possible another flow of energy (input-output). This action may be as simple as a person turning a valve, or it may be the interaction of a limiting fertilizer in photosynthesis.

SELF-MAINTAINING CONSUMER POPULATION symbol represents a combination of 'active storage' and a 'multiplier' by which potential energy stored in one or more sites in a subsystem is fed back to do work on the successful processing and work of that unit.

PURE ENERGY RECEPTOR symbol represents the reception of pure wave energy such as sound, light, and water waves. In this module energy interacts with some cycling material producing an energy-activated state, which then returns to its deactivated state passing energy on to the next step in a chain of processes. The kinetics of this module was first discovered in a reaction of an enzyme with its substrate and is called a Michaelis-Menton reaction.

PLANT POPULATION symbol is a combination of a 'consumer unit' and a 'pure energy receptor'. Energy captured by a cycling receptor unit is passed to self-maintaining unit that also keeps the cycling receptor machinery working, and returns necessary materials to it. The green plant is an example.

ENERGY SOURCE symbol represents a source of energy such as the sun, fossil fuel, or the water from a reservoir. A full description of this source would require supplementary description indicating if the source were constant force, constant flux, or programmed in a particular sequence with, for example, a square wave or sine wave.

(a)

Figure 6
Symbols of the energy-circuit language developed by H. T. Odum (Jansson, 1972)

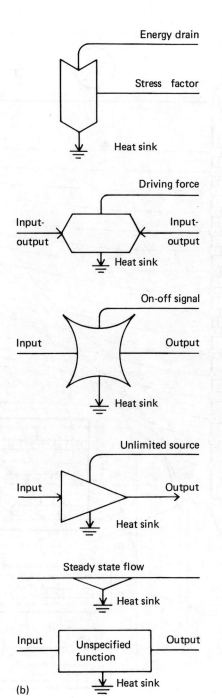

STRESS SYMBOL defines the drain of calories of potential energy flow. When a system is stressed, the potential energy that was available to do work is lost. The curve for a stress factor follows the rectangular hyperbola of a workgate *but* in opposite direction. The stress symbol is then an inverted workgate with energy from the system being drained into a heat sink by an environmental factor (the stress) shown on the opposite side of the workgate.

TWO-WAY GATE OR FORCED DIFFUSION MODULE represents the movement of materials in two directions as in the vertical movement of minerals and plankton in the sea. The movement is in proportion to a concentration gradient of a causal force shown operating the gate. The heat sink shows the action to follow the second law of thermodynamics.

SWITCH is used for flows, which have only on and off states controlling other flows by switching actions. There are many possible switching actions as classified in discussions of digital logic. Some are simple on and off; others are on when two or more energy flows are simultaneously on; some are on when connecting energy flows are off, and so forth. Many actions of complex organisms and man are on and off switching actions such as voting, reproduction and starting the car.

CONSTANT GAIN AMPLIFIER. In this module the amount of energy supplied from the upper flow is that necessary to increase the force expressed by the system by a constant factor, which is called the gain. For example, a species reproducing with 10 offspring has a gain of 10 so long as the energy supplies are more than adequate to maintain this rate of increase.

This symbol is used to represent the energy losses associated with friction and backforces along pathways of energy flow.

BOX symbol is used when an unspecified action is being represented, or when the function is unknown or unimportant to the point being made. If the function is known, but no specific symbol is available, a box may be used with the function written inside.

39

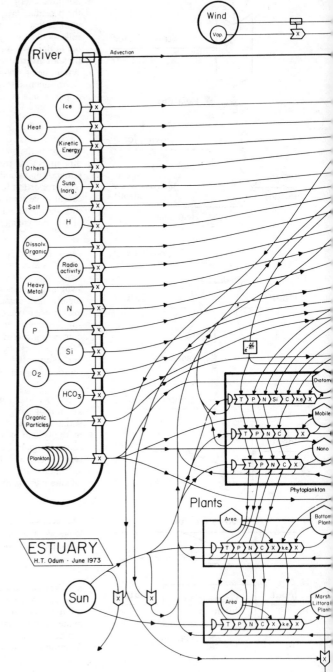

Figure 7
Odum's model of an estuarine system translated
into the energy-circuit language (from Nihoul, 1975)

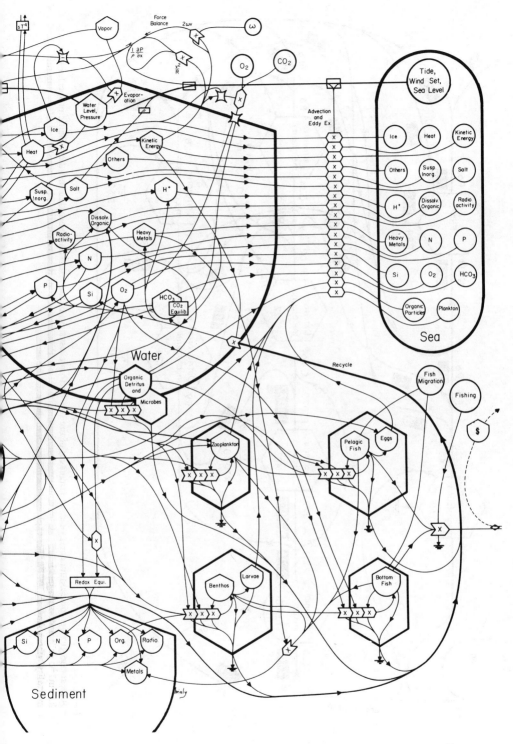

41

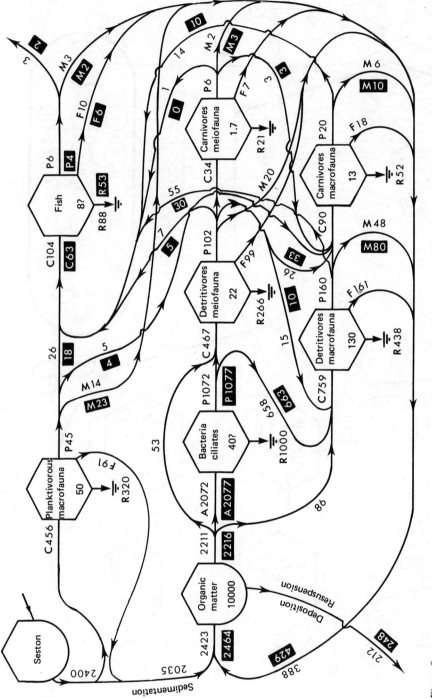

Figure 8
Rough energy flow model (Odum energy-circuit language) of the soft-bottom ecosystem of the Askö-Landsort area (northern Baltic proper). Flows in kJ . m^{-2} . yr^{-1} and storages in kJ . m^{-2}. C = consumption, A = assimilation, P = production, R = respiration, F = egesta and M = 'natural mortality' (from Ankar, 1977). Further explanations are given in the text.

Sensitivity to the parameters

Imagine that a dynamic simulation model has been constructed as described in Section 1.3.1 and tuned such that with the chosen parameter set the model generates numbers, in the time domain, having reasonable fidelity with time series measured in the real world. By how much would the output state variables change for a given change in the values of the rate parameters? We might be interested either because we want to know what would be the effect on the output of an error in the estimate of parameter magnitude, or because we want to know what would be the effect of biological adaptation of the parameter values to changing environmental conditions. Or we might ask how a particular model constructed for a specific geographical site (say a particular estuary) would behave if the rate parameters were all changed to be appropriate to a different particular estuary. In other words, how general, or generalizable, is this particular model? Two main lines of approach are available to the question of sensitivity.

The less elegant or brute-force approach is usually called empirical sensitivity analysis. It consists in running simulations, varying each parameter one at a time, either positively or negatively, and observing the effect on the output state variables. An illustrative example, for a simplified aquatic eco-system is given in Smith (1970). This is a laborious way to proceed, but if the model is extremely complex it may be the only way.

The more elegant approach is usually called analytic sensitivity analysis, of which the classic exposition is by Tomović (1963). The outstanding application of this method to a marine ecological problem is that by Wroblewski and O'Brien (1976) to their spatial model of phytoplankton patchiness. The analysis is done on the non-dimensional form of the equations (see Section 3.2). It is a question of calculating the partial derivatives $\partial V_i/\partial p_k$ of each of the n state variables with respect to each of the m parameters, p_k. This is accomplished by differentiating the steady-state model equations

$$f_i(V_i; p_k) = 0 \qquad (5)$$

with respect to the parameters, p_k, noting that the operator $\partial/\partial p_k$ may be written according to the chain rule as

$$\frac{\partial f_j}{\partial p_k} = \sum_{i=1}^{n} \frac{\partial f_j}{\partial V_i} \frac{\partial V_i}{\partial p_k}; \qquad (6)$$

or in matrix form as

$$\underline{\underline{R}} = \underline{\underline{A}} \cdot \underline{\underline{D}},$$

where $\underline{\underline{A}}$ is an $n \times m$ matrix of the $\partial f_j/\partial V_i$ and R is an $n \times m$ matrix of the $\partial f_j/\partial p_k$. The terms of $\underline{\underline{A}}$ and $\underline{\underline{R}}$ may be calculated about steady-state by use of Equation (5). Since $\underline{\underline{A}}$ is generally non-singular, we may solve for the $n \times m$ matrix of sensitivity coefficients, $\underline{\underline{D}}$, as

$$\underline{\underline{D}} = \underline{\underline{A}}^{-1} \cdot \underline{\underline{R}}.$$

Each term of $\underline{D}$ may subsequently be normalized as

$$\frac{p_k}{V_i}\frac{\partial V_i}{\partial p_k}.$$

Whatever method is used (empirical or analytic) to calculate sensitivity, the sensitivity coefficients may all be further normalized by dividing them by that coefficient with largest value such that a table of *relative* sensitivity coefficients is formed, with absolute magnitudes between zero and unity. Examples of such tables are given in Smith (1970), in Brylinsky (1972), and in Wroblewski and O'Brien (1976). If we are mainly interested in modelling the output of a particular state variable V_i, then an examination of the relative sensitivities $\partial V_i/\partial p_k$ for all parameters p_k will indicate those parameters to which the output of V_i is most sensitive, with the implication that these parameters are the ones that should be determined with the highest precision. Seen in this way, sensitivity analysis can serve as a guide in the allocation of resources to a field programme: it is not worth while to expend much effort to determine a precise estimate of a parameter that has only an extremely small influence on the magnitude of the predicted state variable of most interest.

If the main goal is knowing the relative effect of a given parameter p_k on the suite of state variables V_i (rather than a particular V_i), the relative sensitivities of each p_k can be summed over all variables. In Smith (1970) the absolute values of the relative sensitivities are summed over all variables for each p_k. In Wroblewski and O'Brien (1976), the *ranks* of the unnormalized relative sensitivities are summed. Either way, an index can be established of the relative importance of each parameter to the ecosystem dynamics in general, with implications about necessary precision of estimates as before.

In the sensitivity analysis of their spatial model of phytoplankton patchiness, Wroblewski and O'Brien (1976) found that the largest sensitivity coefficient was that describing the rate of change of nutrient concentration with change in excretion rate of zooplankton. Taking the broader view over all state variables, excretion rate again had the most profound effects on the ecological dynamics. Conversely, the grazing threshold (see Section 1.2) is the parameter with the least influence on the model system.

Construction of a full matrix of sensitivity coefficients can help identify indirect causal effects in the dynamics that might go unnoticed in an intuitive survey of the direct pairwise interactions between variables. Thus a deeper understanding of ecosystem connectivity becomes accessible (Waide and Webster, 1976), rather like that claimed for loop analysis (see Section 2.1.4) but quantitative.

Sensitivity to initial conditions

In non-linear time-dependent prediction models we have to consider also how sensitive will be the output to the values assigned to initial conditions. Whenever the estimates of initial conditions are subject to error, the errors may propagate and grow through successive iterations of the predictive computation until the value of the predicted variable is no more reliable than would be any value chosen randomly from the set of possible values of the state variable (Lorenz, 1969). This sets a fundamental limit to the time in the future for which valid predictions can be made: in their application to the problems of meteorological forecasting these ideas have been advanced furthest by Lorenz and others. The concepts are intimately allied with those of scale (see Section 3.2) in that it can be shown that reducing the magnitude of the error on the initial condition has minimal effect on the estimate of predictability time: errors propagate rapidly at small scale, and reducing the initial error by half merely doubles the time taken for it to reach the next smallest scale, but has no effect on the (slower) propagation of the error variance to all larger scales. The possible application of the notion of predictability time to marine ecological problems is discussed in Platt *et al.* (1977).

Sensitivity of functional form

We may also want to discuss the sensitivity of coupled process models to the exact functional form of one or more of the component functions describing individual processes (see also Section 1.2). For example, in using a model of the light saturation curve for phytoplankton as a component of a complex ecosystem model, it may be sufficient to choose any of a number of possible representations provided that the chosen function is linear at low light intensities and asymptotic to a constant value at high light intensities. Subtleties arising from the exact choice of functional form for this one process may be lost in the noise associated with numerical realizations of complex models. The benefit of this to the modeller is that all functions are not equally easy to deal with in mathematical analysis. For example, Jassby and Platt (1976) concluded that a hyperbolic tangent function was the most consistently successful representation of empirical data on light saturation in phytoplankton. They noted, however, that this function cannot be integrated over depth. For applications involving the vertical dimension explicitly, it is possible to choose another functional form with little loss of precision in the estimates.

Sensitivity to model structure

Few modellers have analysed the effect of alternative descriptions of the basic structure of the system. One of the few such examples is the study by Sjöberg (1977), who showed that the addition of a nutrient recycling loop in the phytoplankton-zooplankton model analysed by Steele (1974) had a great influence on the stability of the model and thus on the conclusions drawn by Steele and others on the behaviour of the real system. The differences between these models are shown in the conceptual diagram in Figure 9. This example may be regarded as sensitivity analysis in the widest sense and leads back to the problems arising when setting up a model. The stabilizing effect of including nutrient

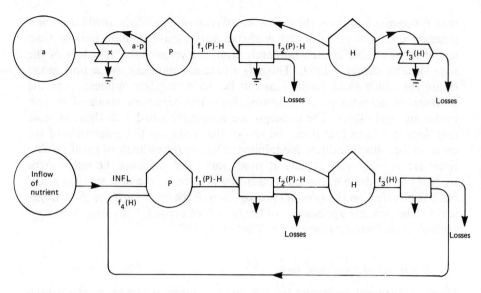

Figure 9
Energy-circuit language models of the phytoplankton (P): zooplankton (H) model analysed by
Steele (1974) (upper scheme) and the modified model analysed by Sjöberg (1977) (lower scheme)
with a nutrient recycling loop added (from Sjöberg, 1977)

recycling was also realized by Steele himself, and discussed by him in Steele
(1977).

1.4.3 VALIDATION[1]

The question of quality control on the modelling process, the evaluation of
the product, is usually referred to as the validation problem. It is an extremely
difficult one.

The validation problem is most acute for complex coupled process models
that generate time series of many state variables. For this kind of output
there is no conventional, objective criterion of goodness-of-fit, even assuming
that there is a suitable data set available for comparison. Model fits are judged
subjectively and for this reason it is usually impossible to judge between
alternative models. For models generating time series of a single variable, an
evaluation index sometimes applied in oceanography is the skill index (a
correlation procedure) of Davis (1976). Calculation of the skill index may
first be done on the same data set used to construct the model (the hindcasting
problem) and then done on a data set from which the model is independent
(the forecasting problem).

Caswell (1976) stresses the importance of exploiting the validation procedure as an opportunity to choose between alternative models or hypotheses.

1. By T. Platt and G. Radach.

We have already noted that this is next to impossible for models simulating many state variables, $V_i(t)$, in the time domain. The problem is compounded by the unfortunate truth that typical data available to ecologists are notoriously noisy. There is some hope of a better future if we can break away from the kind of thinking that considers these time series $V_i(t)$ as the *only* possible outputs from simulation models.

For example we might look at the statistical moments of the $V_i(t)$ series. To take a trivial example, population averages of the validation data set will be more reliable than individual, discrete-time population estimates. Higher-order moments may themselves be invested with a particular significance in ecological theory. For example, Kerner (1962) develops some theory concluding that variance of population size, averaged over all species, is an important ecosystem property (see Section 2.2.4). Looking at the more general analysis of the variance problem, Platt and Denman (1975) have suggested that the time-domain output of simulation models be recast in the frequency domain such that the power spectral representation of the model results be compared against that of the validation data: do the dominant time scales of fluctuation of the model results correspond with those of the data? How do the important phase lags compare in the two systems? (See Section 2.2.5.) Again, we might evaluate a model by comparing the particle-size spectrum implied by it with that observed in the real world (see Section 2.2.3). How would the model particle-size spectrum change under perturbation, say a fivefold increase in nutrient input? Finally, we might compare the sensitivity coefficients of the model with what is known about the sensitivity of the real world system (see Section 1.4.2). If the model is highly sensitive to a parameter in the forcing function that is known to vary by, say, an order of magnitude on a time scale of weeks (e.g. the viscous dissipation rate of turbulent energy in the ocean), there may be a case to make for modification.

A phase-plane representation of time-domain outputs is another possibility. For example, a model may generate time series for phytoplankton and for zooplankton abundances. These outputs will be impossible to test by sampling the real world, given the problem of spatial heterogeneity in the biomass and given that the phase relationships of this dynamic coupled system may also vary from station to station. However, the joint probability distribution of phytoplankton with zooplankton is accessible by sampling and can easily be obtained from the model output (e.g. Steele, 1974).

We suggest then, that the interest of coupled-process models might be extended if their conventional outputs are transformed to derive the macroscopic ecosystem properties discussed at some length in Chapter 2. Not only will this facilitate the selection between alternative hypotheses using noisy data, but in the end it might teach us more about the subtleties of ecosystem behaviour. Questions about the coherence of macroscopic ecosystem properties between model and data probably have a more profound significance than questions about the parallelism between model and data in the time domain.

1.4.4 ESTIMATING RATES BY INVERSE METHODS[1]

The same model system which is used for hindcasting and sensitivity analysis can be used to estimate the constants which appear in the representations of the processes, provided sufficient data on the standing stocks are given. Sufficient means that there should be much more data available than constants to be estimated. These constants, which are prescribed in a hindcast or modified in sensitivity analysis, will be used to fit the model to the measurements. For this purpose the model algorithm will be implemented in an inverse technique, which calculates the constants in the model under the constraint that the error between model results and measurements is minimized by the choice of these constants.

It should be noticed that this procedure is different in quality from, say, polynomial fitting of data, because the terms in the model and the appropriate constants have a biological meaning, which is generally not the case in polynomial fits.

Instead of comparing measured and calculated standing stocks, as usually done in hindcasts, here the consistency of the estimated constants with measured constants has to be checked. Thus the attention has shifted from standing stocks in hindcasting to fluxes in inverse methods.

The method has not yet been widely used in marine biology in connection with coupled process simulation models (e.g. Radach *et al.*, 1979). The method can unfold its full capability only in conjunction with the estimation of the reliability of the calculated constants in relation to data.

1.4.5 COMPARISONS OF COUPLED PROCESS MODELS[1]

Having in mind the subjects of ecosystem research, i.e. the description and demarcation of ecological spaces, description of their component organisms, and the investigation of the functional interrelationships between the organisms with each other and with the non-living environment, we notice immediately that coupled process models can only be useful in a small part of total ecosystem research. This part consists of problems in which a quantification is possible. Non-quantifiable entities in ecosystem research cannot be treated with the help of coupled process models, and there are many examples of these, e.g. communication, adaptation and evolution.

Looking at an interaction diagram it is apparent that only such relations can be modelled as can be expressed in common units, i.e. units of matter or energy. Thereby the variety of possible coupled process models is strongly limited, namely to models of transport of matter or energy.

Since about 1945 mathematical models have developed in marine ecology in ever-increasing numbers and complexity to describe quantitatively the dynamics of primary production in the sea and of the next trophic levels in their dependence on biotic and abiotic factors. In most cases the models were

1. By G. Radach.

developed for the interpretation of one special data set. They served as means to simulate the measured dynamics a posteriori, i.e. were used for tests of 'hindcasting'. Usually time scales of weeks to one year (spring plankton bloom, annual cycle, upwelling events) were investigated, partly under meteorological forcing, often under strong simplification of physical transport processes or their omission. Very often the models yield results which agree with the measurements in the order of magnitude. The deficiencies of the existing models are the following:

1. The models were developed and tested on the basis of one data set only.
2. The complexity·of the models which is *necessary* for the interpretation of the data was not determined. Very often highly complicated process models are incorporated without showing that this degree of complication is necessary to interpret the data.
3. The goodness of the simulation results in relation to the data is not quantified. The judgement about the quality of the simulation is based on visual comparison of measured and calculated curves.
4. Often the *critical* parameters are not compared. In most models comparisons of standing stocks are used for judging the quality. But it seems more sensible to compare the *fluxes*, which lead to the standing stocks. For instance, a small standing stock can be the result of high production and high consumption by a different standing stock. In most cases measurements of fluxes are lacking.
5. The physical transport processes are often too strongly simplified or ignored.

From this the following conclusions can be drawn:

1. Data sets have to be created which are sufficiently complex to allow testing of existing models and their further development. For this purpose the measurements have to be brought into a consistent form, i.e. intercalibration questions regarding measurements with different methods or instruments have to be sorted out. An organization of data adequate to digital data processing has to be provided.
2. A sequence of models, or a model hierarchy, should be investigated instead of one single model. The reason for this is to enable a discrimination between different options of process representations, and to find the simplest model which does not contradict the existing knowledge and which is consistent with the data.
3. The measurement of fluxes should be strongly encouraged.
4. The importance of the physical transport processes has to be checked thoroughly for the given problem. They have to be incorporated in adequate complexity in the biological models.

Although many (partly similar) models were proposed during recent years, at the present time no sufficiently tested model exists, e.g. for the flow of carbon, nitrogen or phosphorus in the lower trophic levels of the food web of the pelagic ecosystem.

From our point of view the growing discredit of coupled process models has its roots in the abovementioned shortcomings and, moreover, in one of the several kinds of misuse which can be encountered:

1. The most frequent misuse consists in forgetting the assumptions of the model and the limits of applicability. Even if the model is validated, the extrapolation of the validity in the future may lead to overinterpretation by inadequate generalization. The deeper reason may be a misunderstanding of the function of a model, which is in fact nothing but a tool.

2. Overlooking the fact that a coupled process model can only be as good a synthesis in total as the single parts are leads to the condemnation of the synthesis instead of an effort to assess the faulty parts and provide better representations of the processes in question.

3. The use of coupled process models which are not sufficiently validated for prediction may not be responsible. Especially in situations where public management authorities have to decide upon unprecedented developments, there is the danger that inadequate conclusions are drawn from the model or that the model is used as an excuse for avoiding thorough scientific considerations.

In all cases it is the erroneous expectations of the model-maker or the model-user that finally lead to the discredit which ecological coupled process models have to suffer.

2

New explorations of ecosystem properties: a holistic view

2.1 Network analyses

2.1.1 INTRODUCTION[1]

Much of the early work in biology can be categorized as either describing the structure or explaining the function of an organ or organism. As we endeavour to imagine holistic theories of entire ecosystems, we find that the historical approach serves as a helpful guide. The fact that we are increasing the scale of observation, relying upon sometimes complicated mathematics to quantify our ideas, does not detract from the necessity of first attempting to describe the structure of the whole community before daring to embark upon an explanation of total system behaviour.

As long as we are willing to aggregate the various parts of an ecosystem into compartments, it follows that the transfers of energy and matter between compartments constitute a network of flows. In this section we present three representative attempts at the mathematical description of networks. It is no accident that common mathematical tools are used in all three approaches. Since interactions are presumed to occur between pairs of compartments, the most convenient display of all possible exchanges is the two-dimensional matrix.

The first subsection narrates the results of introducing economic input–output analysis into ecology. Along with the analytical tools borrowed from economics, the subtle intuition that a flow can directly engender other flows has likewise been imported into ecological thinking. The full application of input–output analysis assumes that one is capable of quantifying the various flows. Unfortunately, it often happens that only qualitative information is available concerning the interactions (e.g. +, −, or 0).

The second subsection portrays the attempt by Isaacs to deal with what he calls 'unstructured' ecosystems. In fact, they have their own structure, but they do not conform to the Eltonian trophic pyramid. Although the mathematical form of this analysis is almost identical to that of input–output analysis,

---

1. By R. E. Ulanowicz.

the economic flow hypothesis is missing. Rather, a flow is more conventionally assumed to be the result of a driving potential (biomass content).

The final subsection, on loop analysis, shows how it may be possible, under certain conditions, to infer the qualitative nature of compound interactions.

2.1.2 ECONOMIC INPUT–OUTPUT ANALYSIS [1]

Economic input–output analysis is attributed to Leontief (1951) and was first introduced to the ecological literature by Hannon (1973). Hannon's discussion of flow analysis was not holistic in scope. Rather he was concerned with the analytical question of how each external input to a system affects any other compartment in the system. In ecological terms one might consider, for example, the case of an estuary receiving energy in the form of both detritus from upstream drainage and production within the estuary from plant producers. A given top carnivore is sustained by these two sources via many interwoven pathways of varying trophic lengths. How does one apportion the effect of these separate inputs on the sustenance of the carnivore? Because the analysis of this mesoscopic question will highlight tools useful in subsequent holistic definitions, it is worth repeating a few concepts and manipulations in Hannon's development.

To begin, the system is considered to be in steady state so that the sum of all inputs to a given compartment is balanced by the sum of all the outputs. Hannon lumps all the inputs into compartment i as the element e_i of an n-dimensional vector. The outputs he differentiates according to whether they flow as useful input to some other compartment, j, in the system (P_{ij}) or whether they are respired as r_i. The stationary balance then becomes

$$e_i = \sum_{j=1}^{n} P_{ij} + r_i. \tag{7}$$

A second assumption is that the production energies of a compartment can be accurately expressed as a linear function of the total direct input to the receiving compartment, i.e.

$$P_{ij} = g_{ij}e_j. \tag{8}$$

In other words, g_{ij} is the fraction of the total diet of species j that originates from source i. Equation (7) becomes

$$e_i = \sum_{j=1}^{n} g_{ij}e_j + r_i, \tag{9}$$

or in matrix form

$$\underline{e} = \underline{G}e + \underline{r} = (\underline{I} - \underline{G})^{-1}\underline{r}, \tag{10}$$

1. By R. E. Ulanowicz.

where I is the identity matrix. Hannon shows that $(I - G)^{-1}$ always exists and calls the elements of this matrix the structural elements of the ecosystem. All information about direct and indirect flow pathways is included in $(I - G)^{-1}$.

Digressing for a moment from the holistic purpose of this section, it is interesting to note some of the suggested applications of the structure matrix and its related forms. Levine (1977), for example, uses the power series expansion of the structure matrix to define an index of extended niche overlap—an extension of Levine's overlap index to include indirect, as well as direct, energy flows. He gives examples of how the nature (sign) of an indirect (i.e. compound or secondary) interaction can be opposite to that of the direct binary relation.

Elsewhere Levine (1980) couples input–output analysis with Markovian statistics to derive a measure of trophic position—the average trophic level at which a species feeds. Trophic position is the first moment (mean) of the pathway length weighted by the probability of the species feeding at that level. The second moment (variance) gives a measure of the trophic specialization of the species.

Ulanowicz and Kemp (1979) proceed in the direction of holism when they use the structure matrix to aggregate portions of populations into trophic compartments. That is, a species feeding along several pathways of different trophic lengths can have its extensive attributes apportioned among several compartments representing different integral trophic levels. The aggregated system, it is argued, will give a more concise picture of trophic dynamics than the complicated diagrams of species flows.

It falls to Finn (1976), however, to construe the structure matrix in a way that it can be used to derive measures of community structure. He begins by defining a measure of the extent of energy flow in the system, the total system throughput (T), as the sum of all the inputs (internal as well as external) to all the species, i.e.

$$T = \sum_{i=1}^{n} e_i. \tag{11}$$

With the help of input-output analysis Finn shows that the average path length $(\bar{L})$ for a given network is simply the total system throughput divided by the total external input to the web, Z.

$$\bar{L} = T/Z \tag{12}$$

Average path length is, then, a measure of the *trophic* complexity in the sense that a more fully developed food web will be characterized by longer food pathways and a larger average path length.

Finn remarked that average path length is strongly dependent on the dimension of the system. He therefore sought an intensive measure, characteristic of the amount of cycling taking place in the system, to compare systems of differing numbers of species. Finn attacked this problem by partitioning the total systems throughput into a straight-through component and a cycled remainder, i.e.

$$T = T_s + T_c, \tag{13}$$

where T_s is the magnitude of the once-through flow and T_c is the amount being cycled. The ratio of the cycled energy to the once-through energy becomes Finn's index of cycling,

$$Y = T_c/T_s. \tag{14}$$

Finn applies his community measures to models of three different ecosystems and provides a comparative discussion. Although he ends the paper by hinting at the potential utility of these indices in testing new hypotheses about ecosystem function, he does not cite what these hypotheses might be.

2.1.3 ANALYTIC FOOD-WEB MODELS[1]

Introduction

The classical concept of Lindeman (1942) envisaged organisms being assigned to definite trophic levels, thus giving rise to a pyramidal structure of the biomass. Much recent research has shown that it is often difficult to assign an organism to a single trophic level and so attention has been switched to a consideration of food-webs. Obviously, for a given number of species there are a large number of possible food-webs, so it is pertinent to inquire if there are any simple or canonical forms of the food-webs that might have general applicability. One such model has been suggested by Isaacs (1972, 1973).

Isaacs's model was prompted by the differences in trace element distributions (mainly caesium) between fish in the Gulf of California and the Salton Sea, an inland salt-water lake. In a simple linear food chain, in which animals only consume others in the trophic level below, it is easy to show that a trace element that is not excreted by organisms will be successively concentrated at each trophic level. If a linear model is assumed and the rate constants are the same at each level, then the ratio of the concentration of the element in the predator to that in the prey should be a constant. It was to test this simple theory that Young (1970) measured caesium, potassium and rubidium concentrations in a wide range of organisms in the Salton Sea. This inland salt-water lake has only been in existence for about seventy years and the food-web of the animals in it appears to be simple. Young found that the caesium concentrations did show the expected increase with position in the trophic ladder. However, when data were obtained from the same species of fish caught in the Gulf of California, this simple picture disappeared. It was found that most of the fish had approximately the same concentration of caesium irrespective of their supposed trophic level of feeding. Also it was found that the

1. By M. J. Fasham.

supposedly primary-feeding mullets had a higher than average concentration of caesium. These results led Young to suggest that the food-web in the Gulf of California, and by inference the open ocean in general, was not pyramidal in form. This in turn led Isaacs (1972, 1973) to investigate in general what he called 'unstructured' food-webs, which will be the main topic of this review. We will first consider the method proposed by Isaacs for studying the problem and will then describe two other formulations which are perhaps more flexible and permit further development.

The Isaacs method

Isaacs assumed that any transfer of matter (or energy) between components in the food-webs can be characterized by three constants, namely:

k_1 = conversion of matter in food to living tissue;

k_2 = conversion of matter in food into irrecoverable forms (e.g. by respiration); and

k_3 = conversion of matter in food into non-living but recoverable forms (e.g. organic detritus),

with the condition $k_1 + k_2 + k_3 = 1$.

He makes the bold assumption that these three constants are the same for all heterotrophic organisms. The time taken for one average step in the food-web is defined as Δt and an amount M_o of autotrophic material is introduced into the food-webs at Δt intervals. Isaacs then investigated the various possible pathways whereby the material M_o can be converted into living and non-living, but retrievable, matter. These pathways can be represented on an infinite matrix (Fig. 10) in which living steps are represented along the columns and recoverable steps along the rows. A diagonal line across the matrix represents the state of the system after a given time. At each point on this diagonal it is possible to count the number of routes whereby living and recoverable matter could reach that point. Viewed along a diagonal these numbers form a binomial series. It is possible to sum them to calculate the total amount of living (M_t^1) and recoverable material (M_t^{11}) at time $t = n\Delta t$ as

$$M_t^1 = M_o k_1 (k_1 + k_3)^n \tag{15}$$

and
$$M_t^{11} = M_o k_3 (k_1 + k_3)^n. \tag{16}$$

These expressions obviously form a geometric series and as $k_1 + k_3 < 1$, they can be summed to infinity to give

$$M^1 = M_o k_1/(1 - (k_1 + k_3)) = M_o k_1/k_2, \tag{17}$$

and
$$M^{11} = M_o k_3/(1 - (k_1 + k_3)) = M_o k_3/k_2. \tag{18}$$

These quantities represent the steady-state values obtained when a constant flux of material equal to $M_o/\Delta t$ is fed into the system. The matter lost to the system is given by

$$M_o - M^1 - M^{11} = M_o(1 - (k_1 + k_3))/k_2 = M_o. \tag{19}$$

55

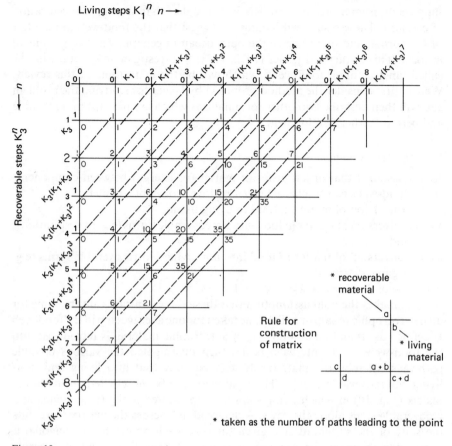

Figure 10
Matrix representing principal characteristics of a generalized food-web (from Isaacs, 1972)

Thus, when the system is in equilibrium, the input flux of plant material exactly balances the respiration.

Isaacs goes on to consider a slightly more complicated food-web consisting of herbivores (M_m), detrital feeders (M_d) and full predators (M_p). It is obvious that

$$M^1 = M_m + M_d + M_p, \tag{20}$$

and it can be shown that

$$M_m = M_o k_1,$$
$$M_d = M_o k_1 k_3 / k_2,$$
$$M_p = M_o k_1^2 / k_2. \tag{21}$$

In all these equations the transfer of food has been implied, but similar equations can be written for the transfer of any trace material, although of course the

56

values of the constants will probably be different. Isaacs used Young's results on the caesium–potassium ratios in the Salton Sea and Gulf of California to attempt to estimate the values of k_1 and k_3 for food. The result is a quadratic equation in k_1 and k_3 and it was shown that assuming a range of reasonable values of k_3 leads to estimates of k_1 which are within the range of conversion values commonly found in laboratory experiments. Isaacs concludes that 'the assumption of an unstructured food-web in the Gulf of California in explanation of Young's findings on predaceous fish leads to reasonable estimates of the coeffecent of conversion of food'.

Matrix formulation of food-webs

It was pointed out by Lange and Hurley (1975) that Isaacs's method could not be applied to the case where, say, different coefficients were assumed for the conversion of dead and living material to irrecoverable forms. To overcome this problem they formulated the model in terms of matrix operations. The matter in the system is represented as a column vector

$$\underline{W} = \begin{bmatrix} W_1 \\ W_2 \\ W_3 \end{bmatrix}$$

where:
W_1 = amount of living autotrophic material in the system;
W_2 = amount of living heterotropic material; and
W_3 = amount of non-living but recoverable material.
They then showed that the passage of material from one component to the next in one time-step can be represented by the matrix operator

$$\underline{\underline{A}} = \begin{bmatrix} 1 & 0 & 0 \\ k_1 & k_1 & k_1 \\ k_3 & k_3 & k_3 \end{bmatrix} \tag{22}$$

acting on the vector W. Thus the state of the vector W after n time-steps will be given by

$$\underline{W}_n = \underline{\underline{A}}^n \underline{W}_o, \tag{23}$$

where $\underline{W}_o$ is the initial state. The equilibrium state will then be given by the limit of $\underline{W}_n$ when n goes to infinity.

This limit can be calculated from the spectral description of the matrix $\underline{\underline{A}}$ given by

$$\underline{\underline{A}} = \underline{P} \underline{\underline{A}} \underline{Q}^T \tag{24}$$

where P and Q are the left and right latent vectors and the elements of diagonal matrix $\underline{\underline{A}}$ are given by

57

$$\delta_i = \lambda_i d_i, \tag{25}$$

where λ_i is the ith latent root and $d_i = \underline{P}_i^T \underline{Q}_i$. In their derivation Lange and Hurley incorrectly used the spectral decomposition for a symmetric matrix, although in fact this error does not effect their final conclusion. It is a well-known result that

$$\underline{\underline{A}}^n = \underline{\underline{P}}\underline{\underline{\Delta}}^n\underline{\underline{Q}}^T \tag{26}$$

and so the state of the system at time $N\Delta T$ will be

$$\underline{W}_n = \underline{\underline{P}}\underline{\underline{\Delta}}^n\underline{\underline{Q}}^T\underline{W}_o. \tag{27}$$

It is always possible to write the initial state $\underline{W}_o$ as a weighted sum of the right latent vector giving

$$\underline{W}_o = \underline{\underline{Q}}^T\underline{C} \text{ where } \underline{C} = \begin{bmatrix} C_1 \\ C_2 \\ C_3 \end{bmatrix}. \tag{28}$$

Substituting Equation (28) in (27) we obtain

$$\underline{W}_n = \underline{\underline{P}}\underline{\underline{\Delta}}^n\underline{C}. \tag{29}$$

For the matrix $\underline{\underline{A}}$ defined by Equation (22) the latent roots are

$$\lambda_1 = 1, \lambda_2 = k_1 + k_3, \lambda_3 = 0, \tag{30}$$

while the matrices P and Q are

$$\underline{\underline{P}} = \begin{bmatrix} 1 & 0 & 0 \\ k_1/k_2 & k_1/k_2 & 1 \\ k_3/k_2 & k_3/k_2 & -1 \end{bmatrix} \qquad \underline{\underline{Q}} = \begin{bmatrix} 1 & 1 & 0 \\ 0 & -k_2/(k_1 + k_3) & -k_3/k_1 \\ 0 & -k_2/(k_1 + k_3) & 1 \end{bmatrix}$$

whence

$$\underline{\underline{\Delta}} = \begin{bmatrix} 1 & 0 & 0 \\ 0 & -(k_1 + k_3) & 0 \\ 0 & 0 & 0 \end{bmatrix}.$$

Using these relationships we find that

$$\underline{W}_n = C_1\underline{P}_1 + C_2[-(k_1 + k_3)]^{2n}\underline{P}_2 \tag{31}$$

where $\underline{P}_1, \underline{P}_2$ are the first two column vectors of $\underline{\underline{P}}$. As n goes to infinity the second term disappears leaving

$$\underline{W}_\infty = C_1\underline{P}_1. \tag{32}$$

The quantities M^1 and M^{11} can now be identified with the second and third row elements of $\underline{W}_\infty$ giving

$$M^1 = C_1 k_1/k_2, \quad M^{11} = C_1 k_3/k_2. \tag{33}$$

58

The initial conditions imply that $C_1 = M_o$ and thus the Lange-Hurley method produces the same result as the Isaacs method.

Lange and Hurley also use this method to derive values for m^1 and m^{11}, when the transfer of material is given by the more general matrix

$$\underline{A} = \begin{bmatrix} 1 & 0 & 0 \\ k_1 & k_4 & k_7 \\ k_2 & k_5 & k_8 \end{bmatrix}. \tag{34}$$

The derived equations are

$$M^1 = \frac{[-k_1(k_8 - 1) + k_2 k_7] M_o}{(k_4 - 1)(k_8 - 1) - k_5 k_7}$$

and

$$M^{11} = \frac{[-k_2(k_4 - 1) + k_5 k_1] M_o}{(k_4 - 1)(k_8 - 1) - k_5 k_7}. \tag{35}$$

Formulation in terms of compartment models

We have seen that the matrix formulation of Lange and Hurley provides a more straghtforward way of calculating the equilibrium values of M^1 and M^{11} and is also capable of being generalized for the case where the constants k_1, k_2 and k_3 will differ depending on whether the conversion is from source, living or dead recoverable material. However, so far, only steady-state values have been derived whereas in practice it may be important to determine the time behaviour of a system. Isaacs touches briefly on this point in his 1973 paper.

Probably the most straightforward way to determine the time behaviour is to formulate the food-web in terms of the familiar compartment model used in tracer kinetics (e.g. Mulholland and Simms, 1976). Consider a subset of an ecosystem consisting of n compartments with compartment i having biomass W_i. Let the flow of biomass from compartment j to i be given by F_{ij}, then the mass balance equation from compartment i is given by

$$\dot{W}_i = \sum_{j=0}^{n} F_{ij} - \sum_{j=0}^{n} F_{ji}, \tag{36}$$

where F_{i0} and F_{0i} represent interactions with the surrounding system. The simplest assumption about the F_{ij} is that they are a linear function of the W_i values and are donor-compartment-controlled, i.e.

$$F_{ij} = a_{ij} W_j \tag{37}$$

for all j. The flows F_{i0} are the input to the subsystem enabling it to reach a steady state. Substituting for the F_{ij} in Equation (36) we obtain

$$\dot{W}_i = F_{i0} + \sum_{j=1}^{n} a_{ij} W_j - \left(a_{0i} + \sum_{j=1}^{n} a_{ji} \right) W_i. \tag{38}$$

It is easy to see that this equation can be written in matrix form as

$$\dot{\underline{W}} = \underline{\underline{C}}\underline{W} + \underline{F} \qquad (39)$$

where the elements of the matrix $\underline{\underline{C}}$ are given by

$$c_{ii} = -a_{0i} - \sum_{j=1}^{n} a_{ji},$$

and

$$c_{ij} = a_{ij}, \ i \neq j.$$

Thus the steady-state values will be given by the matrix equation

$$\underline{\underline{C}}\underline{W} + \underline{F} = 0. \qquad (40)$$

Let us now apply the method to the system shown in Figure 11 which is a compartmental realization of the simplest food-web considered by Isaacs. The three compartments represent the three divisions of the organic matter into plant, living material and detritus. The only input into the system is the flow Q representing the rate of primary production, while matter is lost from each compartment due to respiratory processes, etc. The rate coefficients for transfer between components are shown on the diagram and using these the matrix $\underline{\underline{C}}$ can be written as

$$\underline{\underline{C}} = \begin{bmatrix} -(\alpha_1 + \alpha_2 + \alpha_3) & 0 & 0 \\ \alpha_1 & -(\alpha_2 + \alpha_3) & \alpha_1 \\ \alpha_3 & \alpha_3 & -(\alpha_2 + \alpha_1) \end{bmatrix}$$

while

$$F = \begin{bmatrix} Q \\ 0 \\ 0 \end{bmatrix}.$$

Using these expressions for $\underline{\underline{C}}$ and $\underline{F}$ we can solve Equation (39) to give the following values for the steady-state biomass in each compartment:

$$W_1 = Q/(\alpha_1 + \alpha_2 + \alpha_3);$$

$$W_2 = Q\alpha_1/\alpha_2(\alpha_1 + \alpha_2 + \alpha_3);$$

and

$$W_3 = Q\alpha_3/\alpha_2(\alpha_1 + \alpha_2 + \alpha_3). \qquad (41)$$

If we now make the substitutions

$$\alpha_i = k_i/\Delta t, \quad i = 1, 2, 3 \qquad (42)$$

and

$$Q = M_o/\Delta t,$$

then we obtain

$$W_1 = M_o;$$

$$W_2 = M_o k_1/k_2;$$

and

$$W_3 = M_o k_3/k_2 \qquad (43)$$

which agrees with the previous formulations.

60

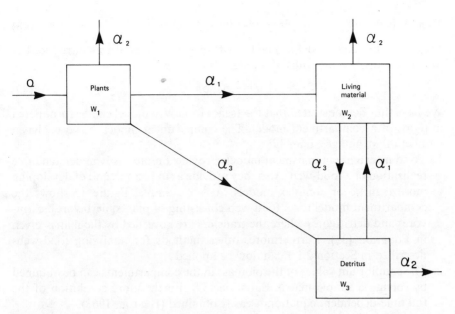

Figure 11
Compartment model of a simple food-web as represented by Equations (41)

In order to justify these substitutions we must first rewrite Equation (38) as a difference equation. Writing $\dot{\underline{W}} = \delta\underline{W}/\delta t$ we obtain

$$\delta\underline{W} = \delta t \underline{\underline{C}}\,\underline{W} + \delta t\underline{F}. \tag{44}$$

If we put 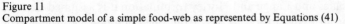 $\delta\underline{W} = \underline{W}^1 - \underline{W}$

we find that $\qquad \underline{W}^1 = (\delta t\underline{\underline{C}} + \underline{\underline{I}})\underline{W} + \delta t\underline{F}, \tag{45}$

or $\qquad \underline{W}^1 = \underline{\underline{D}}\,\underline{W} + \underline{G}.$

This equation is now in the same form as the matrix formulation given by Lange and Hurley.

For the system being considered the matrix $\underline{\underline{D}}$ is given by

$$\underline{\underline{D}} = \begin{bmatrix} 1 - (\alpha_1 + \alpha_2 + \alpha_3)\delta t & 0 & 0 \\ \alpha_2\delta t & 1 - (\alpha_2 + \alpha_3)\delta t & \alpha_1\delta t \\ \alpha_3\delta t & \alpha_3\delta t & 1 - (\alpha_2 + \alpha_2)\delta t \end{bmatrix}. \tag{46}$$

If we compare the last two rows of matrix $\underline{\underline{D}}$ with those of the matrix $\underline{\underline{A}}$ given in Equation (22) it is obvious by inspection that

$$\alpha_1\delta t = k_1 \quad \text{and} \quad \alpha_3\delta t = k_3. \tag{47}$$

By comparing the centre diagonal term it can be seen that

$$k_1 = 1 - (\alpha_2 + \alpha_3)\delta t.$$

Therefore $\qquad\qquad \alpha_2 \delta t = 1 - k_1 - k_3 = k_2.$ (48)

In Isaacs's original model a quantity M_o of plant material was introduced at time intervals Δt apart; thus, if we identify $\Delta t = \delta t$, then

$$Q = M_o/\delta t.$$ (49)

We have thus demonstrated that the Isaacs food-web model can be formulated in terms of a compartment model. The compartment model, however, has a number of advantages, namely:

1. To some extent compartment models provide a more easily understandable realization of a food-web. Also they provide a simpler method of developing more realistic or complex models. As an example, Figure 12 shows the compartment model for a food-web consisting of plants, herbivores, carnivores and detritivores where the transfers are governed by the matrix given in Equation (33). Furthermore, other methods for analysing food-webs described in Section 2.1.1 can now be applied.

2. The equilibrium values of the biomass in the compartments can be obtained by solving a simple matrix Equation (39). Furthermore a solution of the full time-dependent equation is easily obtained (Hearon, 1963).

3. In both the Isaacs and Lange and Hurley formulation it is assumed that the time taken for an item of food to pass from one component to the next is the same irrespective of the path. Isaacs (1972) discusses how this restriction

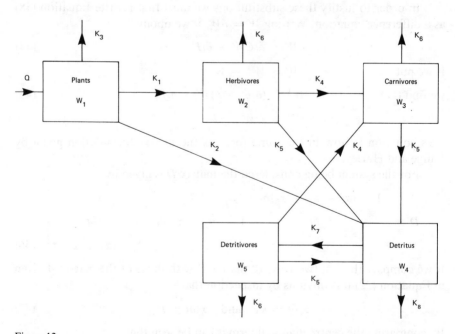

Figure 12
Compartment model of a more complicated food-web, corresponding to Equations (33)

can be relaxed for the most simple case, but this generalization is automatically catered for in the compartment model as the a_{ij} terms are defined as rate constants.

4. The compartment model is easily generalizable to allow for nonlinear flow rates.

Some deductions from the generalized food-web model

In the normal linear hierarchical food-web in which herbivores eat plants and are themselves eaten by carnivores, the biomass of the carnivores must always be less than the herbivores. For an Isaacs food-web, however, this need not be so. From Equations (21) the biomass of herbivores and predators is respectively

$$M_h = M_o k_1 ;$$

and
$$M_p = M_o k_1{}^2 / k_2 . \tag{50}$$

Thus the condition for $M_p > M_h$ is given by

$$k_1 > k_2 \quad \text{or} \quad 2k_1 > 1 - k_3 . \tag{51}$$

Isaacs (1973) considers the case $k_3 = 0.3$ for which the condition that $M_p > M_h$ is $k_1 > 0.35$. He suggests that these values are not unreasonable and may explain the results of a number of pelagic field studies in which the biomass of predators was surprisingly high. Another area where this theory might have an application is in the deep-sea benthos, where recent results have suggested that the biomass of the large megafauna is of the same order of magnitude as the macrofauna on which it supposedly feeds (Haedrich and Rowe, 1977). It was suggested by these authors that this result highlighted the importance of large lumps as a food supply, but an Isaacs food-web might be a reasonable alternative explanation.

Conclusions

It has been shown that the 'unstructured' food-web model suggested by Isaacs can be considered in a more general context by the use of compartment models. It can then be seen that there will be a whole range of food-webs ranging from the more aggregated models of Isaacs to large complex models where each species is considered as a separate compartment. Exactly where along this spectrum a valid description of a marine ecosystem can be found is not yet apparent. If the simple food-web of Isaacs is valid then this has important consequences for marine ecology in that predicted levels of biomass, trace elements or pollutants will be very different from those predicted by a trophic level model. The evidence for the Isaacs food-web comes from the results of just one experiment, although some supporting evidence for benthic ecosystems has been provided by Hargrave and Phillips (1976). The processes whereby marine animals accumulate trace elements are extremely complex (Pentreath, 1977) and it is therefore essential that further carefully planned experimental work be carried out to provide further evidence.

2.1.4 LOOP ANALYSIS[1]

It is a truism that even the simplest biological systems are far too complex for us to specify in complete detail, such as is possible for some mechanical systems. Loop analysis is a technique that recognizes at the outset that the system under study will be only partially specified: it seeks only qualitative conclusions. A good introduction is the paper by Levins (1974).

An important application is that of Lane and Levins (1977) treating the effect of nutrient enrichment on model plankton communities. The general thesis of this paper is that results of physiological experiments on single species cannot necessarily be extrapolated to determine the response to perturbation of that same species as part of an ecological community, because of the plethora of other interactions which might confound the simple response isolated in the laboratory.

In loop analysis, we assume that the algebraic sign of all these interactions is known, but not necessarily their intensity. The system may then be represented as a network diagram, with a state variable at each node or vertex of the network. The analysis stands or falls depending on the adequacy of the network diagram as a representation of the system. The system is only partially specified in that we do not quantify the strength of the interactions, just their directions. The signs of the indirect dependencies (as opposed to the direct interactions specified on the network diagram) can sometimes, but not always, be deduced without ambiguity. In the case that a given interaction is non-linear, its sign may change as the magnitude of the independent variable increases. Then, two diagrams would be necessary, each one applicable to the appropriate range of the variables.

A recent, apparently successful application of loop analytical methods was given by Briand and McCauley (1978), for a lake ecosystem. The authors assumed that the plankton system of the eutrophic lake could be described with a network of six state variables: nutrients, inedible algae, edible algae, herbivores, carnivores and planktivorous predators. A table was constructed showing the predictions concerning the direction of changes in the state variables following perturbations of each of the six state variables taken separately. The authors provided persuasive evidence, based on their own manipulations of the lake, and on studies published by other workers on different lakes, showing that the expected changes are consistent with what has been observed to happen in the real world. It is stressed that often the changes predicted by the analysis of the network are opposite to what might be predicted from a

1. By T. Platt.

knowledge of the nature of the pairwise interactions of the state variables taken two at a time.

There are no comparable studies available for the marine environment. Two possible criticisms might be levelled against the work. One is an apparent confusion between nutrient concentrations and nutrient rates of supply. The other is that although loop analysis deals with systems near equilibrium, the study is intended to deal with a lake driven to produce toxic algal blooms through perturbation of the nutrient fields. Is this equilibrium or not? The authors do say explicitly that all stability requirements are satisfied for their model, and their results are sufficiently intriguing that the study should be taken seriously.

2.2 Thermodynamics and statistical mechanics

2.2.1 ECOLOGICAL APPLICATIONS OF IRREVERSIBLE THERMODYNAMICS[1]

In thermodynamic terms, ecosystems are open and dissipative. That is, they are maintained in a state away from equilibrium (= all organisms dead) by a continuous throughput, modification and dissipation of energy (sunshine). For open and dissipative physical systems, a considerable body of theory (irreversible thermodynamics) has been developed and applied with some success to such questions as 'How does the efficiency of system operation depend on the rate at which energy is dissipated by it?' (Odum and Pinkerton, 1955).

The questions at issue here are 'Can this body of theory be applied to ecological problems?' and, if so, 'What can it tell us about the general principles upon which ecosystems are constructed?' A fundamental problem is that much of irreversible thermodynamics deals with systems close to their equilibrium configurations. Does this constitute a prima facie case for its invalidation in ecological applications?

For any system we can write a general equation enumerating the various ways in which power is dissipated in the system. Their sum is equal to the rate of entropy production.

$$T\frac{dS}{dt} = \sum_i J_i X_i.$$ (52)

Here S is the entropy, T the absolute temperature and the products $J_i X_i$ each have the dimensions of power. The implication of writing the power dissipation terms in this way is that the processes represented by each of the terms can be regarded as a product of a generalized flux J and a thermodynamic force X. An example would be the flow of liquid along a pressure gradient.

1. By T. Platt.

One can further write an expression for the J's as

$$J_i = \sum_j L_{ji} X_j. \tag{53}$$

The implication here is that the various forces are coupled to each other, and that altering the magnitude of one force might modify the flux associated with another force. The L_{ij}'s are the coupling coefficients, known as the phenomenological coefficients. A second implication of Equation (53) is that the fluxes are linearly related to the forces. This restriction will probably turn out to be unacceptable for the classes of problems we are concerned with in ecological theory.

We are guided in the application of Equation (52) and Equation (53) by minimization principles. Thus, for near-equilibrium conditions, the criterion for a stable steady state is one of minimum entropy production:

$$d(\sum_i J_i X_i) = 0. \tag{54}$$

But our thermodynamic picture of the ecosystem is of a system driven far from true equilibrium by continuous dissipation of energy. That is, we look for a stable state different from that described by Equation (54). This more general criterion, according to Prigogine, is

$$\sum_i J_i dX_i = 0. \tag{55}$$

Ulanowicz (1972) has discussed how an ecosystem description might be facilitated by these principles. One conclusion of the analysis is that the specific energy content of the organisms should increase in the sequence: primary producers, herbivores, carnivores.

Odum and Pinkerton (1955) have discussed the structure of biological systems in terms of irreversible thermodynamics with a view to finding the criteria for optimal efficiency of operation. Using the formalism of Equation (52) and Equation (53) and a hypothesis due to Lotka that natural systems will evolve according to selection for maximum power output, they deduce that the efficiency at maximum power output can never exceed 50 per cent. (In fact this figure of 50 per cent is in error: Smith, 1976, gives the correct answer as 61·8 per cent.) Their discussion illustrates the general point that, in applying physical arguments to biological problems, we are rarely sure what the optimization principles involved are. Are ecosystems, populations, individuals designed to operate at maximum efficiency? At maximum growth rate? Or what?

The other basic problem here is the identification of the thermodynamic forces X_i leading to the biological fluxes J_i. This is a non-trivial problem. Indeed, it is quite possible that analogues of the X_i do not exist for ecological problems. The earliest attempt to find an ecological force is probably that of Odum (1959). The corresponding flux involved is the biomass flux through the food chain. Odum thought that this flux had to be related to the concentration of food, by analogy with the correspondence between rates of chemical reaction and concentrations of reactants, but was not able to bring the matter to a firm conclusion.

Phenomenological arguments have been pushed furthest for problems of growth and development by Zotin (e.g. Zotin, 1973). His approach is to start from the premise that relevant phenomenological coefficients exist and to solve for the coefficients by substituting in the relevant process equation. As an example, let us consider the treatment for the individual growth equation. Define J_g to be the flux for change in individual biomass B.

$$J_g \equiv \frac{1}{B} \cdot \frac{dB}{dt}. \tag{56}$$

Zotin uses the Von Bertalanffy growth equation

$$\frac{1}{B} \frac{dB}{dt} = k_{gg} \left(\frac{B_m^b}{B^b} - 1 \right) \tag{57}$$

where k_{gg} and b are constants and B_m is some maximum weight. Then comparing Equation (56) and Equation (57) with the definition Equation (53) we can see that X_g, the force leading to the growth flux J_g, is

$$X_g = r_g \left(\frac{B_m^b}{B^b} - 1 \right), \tag{58}$$

where r_g is simply a constant to ensure dimensional consistency.

So far such developments are little more than the construction of a formalism: no direct applications have been made. One is tempted to believe that ultimately the way will be clear to its exploitation in ecological theory, since ecosystems have to conform to the law of conservation of energy. But we lack a picture of what the biological forces are that cause the changes we see, at all levels of complexity above that of the biological membrane.

2.2.2 INFORMATION THEORY APPLIED TO ECOSYSTEM STRUCTURE[1]

Since biological thermodynamic forces are so obscure, it may be worth exploring how far one may go in describing community behaviour without having to invoke forces. Indeed, it turns out that limiting discussion to fluxes alone allows the application of information theory to network flow structure in a manner which is more satisfying than the *ad hoc* acceptance of species diversity indices. Thus, apart from some arbitrary identification of system compartments, the remainder of this section will be devoted to analysing natural communities as though they consisted exclusively of fluxes.

---

1. By R. E. Ulanowicz, Contribution No. 967 of the University of Maryland, Centre for Environmental and Estuarine Studies.

Section 2.1.2 has already introduced the reader to some current efforts to analyse systems purely on the basis of flows. There it was mentioned that an appropriate scale for a compartment was the total amount of flow passing through it. The sum of the individual throughputs, in turn, defined a community scale factor, the total systems throughput (T). Growth of community scale is generally regarded as an increase in T (e.g. the familiar growth in gross national product).

However, bigger is not necessarily better, and one desires a measure of organization within the community. For the purposes of this discussion organization will be taken to mean the tendency for the community to behave as a coherent whole, as distinct from a mere collection of independent parts. Now for two compartments to act in consort with one another implies that a pathway of communication exists between them. This communication can be achieved by an infinite variety of mechanisms; however, the only interchange visible to an observer who can sense only fluxes will occur along flow pathways. Information can be sent in the form of the magnitude of a steady flux or as a temporal variation in the flow. As in the earlier section on input–output analysis, it will be assumed that the system is at steady state, thereby limiting consideration to steady flows. This assumption is not as restrictive as it may first appear, as choosing an appropriate time interval can cause most dynamic systems to appear to be in quasi steady state.

It is worth mentioning at this point that the pathways of communication and sustenance are perceived to be identical. For species i to communicate with species j implies that i sustains in part the perceptible existence of j (the magnitude of j being defined by its throughput). Attention is now focused on those situations where there is mutual sustenance among compartments. In such circumstances a positive feedback of sustenance occurs wherein the growth of any participating unit will be reinforced by the mutual exchange. If an arbitrary collection of compartmental throughputs is subject to conditions propitious to their growth, those subdivisions engaged in mutual sustenance will grow at a faster rate than the isolated compartments and will emerge as the dominant elements in the community. Such positive feedback phenomena are fundamental to the theory of cybernetics and their application to eco-systems has been emphasized by Margalef (1968) and Odum (1971).

Most of the cited discussions have been directed towards reward loops and attention seems concerned with effects of such feedback upon the individual compartments. But to entertain a holistic view of the system implies that one should search for a means of characterizing the mutual sustenance intrinsic to the entire community. Since the networks of internal fluxes and com-munication have been assumed identical, it is now appropriate to borrow from information theory to define the average mutual sustenance.

Following Mulholland (1975), the fraction of the total systems throughput pertaining to compartment i will be labelled Q_i. Also, that fraction of species i throughput which flows to directly sustain species j is denoted by f_{ij}. The average mutual sustenance (taken from the average mutual information) is defined as

$$A = K\sum_j\sum_k f_{kj}Q_k \ln(f_{kj}/[\sum_i f_{ij}Q_i]) \tag{59}$$

where K is an appropriate scalar. Now the most appropriate scale for the system has already been assumed to be the total systems throughput. Setting $K = T$ and calling the scaled quantity the ascendancy, it seems probable that during an increase of system size the pattern of flows most likely to emerge would be the one with greatest ascendancy.

It is important to realize that ascendancy cannot increase without bound. One limiting factor is immediately seen to be the external constraints of finite input rates. But input rates alone do not stop ascendancy from increasing, since ever-faster internal cycling of medium (e.g. free energy, mass) could allow ascendancy limitless value. It is immediately obvious, however, that infinitely high internal cycling is proscribed by the second law of thermodynamics. Each pass through a compartment must result in a finite rate of loss of medium. Odum and Pinkerton (1955) contend that near equilibrium, the useful power output of any stage is maximized when the efficiency (defined as the rate of power input to the rate of generation of useful power) is 50 per cent. This is a conservative result, and one has every reason to expect that far-from-equilibrium (e.g. ecological) processes will deliver maximum output at smaller power efficiencies.

Hence maximization of ascendancy is seen to be a compromise between maximizing the scale of the system and achieving the greatest mutual sustenance. Beyond an optimal point, 'attempts' by the system to increase total throughput result in lowered efficiencies and a more than compensating decrease in mutual sustenance. Conversely, for a fixed community of species to increase mutual sustenance by becoming more efficient will decrease T by an even greater proportion.

Now because a community with greatest ascendancy is the one most likely to emerge during growth does not imply that the same configuration will persist. To study the question of persistence it is necessary to consider other components of the total system complexity.

By scaling the familiar Shannon Index by the total throughput, one obtains an expression for the system development capacity, i.e.

$$C = -T\sum_i Q_i \ln Q_i. \tag{60}$$

Information theory allows the partitioning of this index into four relevant components:

$$C = A + A_o + S + R \tag{61}$$

where the first of these, the ascendancy, has already been defined.

The second component, A_o, is the uncertainty ascribed to the flows of medium exported from the community, i.e.

69

$$A_o = -T\sum_i f_{ie}Q_i \ln(Q_i) \qquad (62)$$

where the subscript e refers to the external universe.

The third component, S, defines the uncertainty encumbered by necessary dissipation.

$$S = -T\sum_i r_iQ_i \ln Q_i \qquad (63)$$

where r_i is the fraction of throughput of compartment i that is respired.

The one remaining component, R, is termed the redundancy,

$$R = -T\sum_i\sum_j f_{ij}Q_i \ln(f_{ij}Q_i/\sum_k f_{kj}Q_k) \qquad (64)$$

because it is a measure attached to the redundancy of the flow pathways within the community.

Rewriting Equation (61) as

$$A = C - (A_o + S + R) \qquad (65)$$

one can easily recognize the similarity to the Gibbs or Helmholtz free energies. The development capacity, C, is analogous to the internal energy of a thermodynamic system. Like internal energy, not all of the development capacity is accessible to the system. There will always remain an amount $(A_o + S + R)$ as necessary overhead to *maintain* the system configuration.

The central hypothesis of this section can now be stated: 'A self-organizing community always behaves to optimize its ascendancy subject to hierarchical, environmental, and thermodynamic constraints.'

It is useful to consider how the value of ascendancy might increase in ecosystems through changes in the remaining four state variables. Where helpful, examples of possible mechanisms for the change will be mentioned.

An upper bound on the ascendancy is the development capacity, C. Increase in C allows more possibility for A to rise. Thus, increases in the scale of the system occasioned by increased input fluxes or increasing the partitioning of flow among more species will both augment C. In reality there are limits to how finely flows can be partitioned, since very small populations are more susceptible to chance extinctions. Therefore, availability of resources and environmental perturbations determine that C shall have an upper bound. (What that bound might be is a more complicated issue, since T can be influenced by mutual sustenance.)

Ascendancy is prevented from achieving full development capacity by the overhead inherent in the sum $(A_o + S + R)$. One is tempted at this point to discuss how this sum might be minimized, until it is realized that all the variables are non-conserved (Denbigh, 1975). It is possible to increase ascendancy by increasing the scale of the system, so that the value of the difference between development capacity and overhead can increase, even if overhead is becoming a progressively larger fraction of total development capacity.

Ecosystems do not always develop so as to minimize overhead, in fact the reverse circumstance is probably more common.

If input and output fluxes are completely independent, then minimization of A_o appears a sound strategy for system development. If, however, the given system appears as a compartment in a higher-level system, then exports can become a crucial contribution to the ascendancy of the larger realm (and thereby the continued existence of the smaller community).

In similar manner, if environmental variability is low, the system can develop by reducing redundancy to the point where specialization becomes a prominent feature of the network (e.g. the tropical rain forest). However, such a strategy would be inappropriate under conditions where environmental variation is great. There, network redundancy provides increased freedom for the system to reconfigure itself so as to maintain the level of ascendancy. This is the crux of Odum's (1953) argument that pathway redundancy facilitates system homoeostasis.

The least clearly defined component of the overhead is the dissipation measure, S. The ambiguity stems from the lack of a concrete expression for the optimal efficiency for maximum power production in the far-from-equilibrium realm. Indications are that the optimal efficiency falls monotonically with increasing output, but this notion needs quantitative refinement. The qualitative hypothesis of falling efficiency with increasing power output makes it likely that S will increase during the course of development. The possibility cannot be excluded that under particular conditions the system might develop so as to decrease S, but such conditions represent a (probably small) subset of general circumstances.

No assumption has been made concerning the distribution of the compartments in space. For the sake of simplicity the reader has probably assumed that each compartment represents a species homogeneous in space. But spatial structure and flows are quite important, especially in marine ecosystems. Fluxes have not been constrained to those resulting in biomass transformation, but may also include those associated with biomass transport. Thus, if one is interested in n biological entities distributed over m segments of a spatial domain, there is no conceptual difficulty in treating the generalized flow network among $m \times n$ compartments with the development hypothesis. Maximal ascendancy over both space and species should describe both spatial and trophic structure. Input and output uncertainties (e.g. A_o) are likely to be dominant in marine ecosystems.

The development hypothesis cannot be accepted as valid unless empirical work shows that the macroscopic descriptions it engenders are applicable to a large body of general observations. There are indications that many of the qualitative remarks about ecosystem development (E. P. Odum, 1969; H. T. Odum, 1971) are consistent with the hypothesis (Ulanowicz, 1980), but further investigation is obviously necessary.

Nevertheless, the very possibility of a development criterion based solely on fluxes could have serious implications on the future course of marine ecosystems modelling. Simulation modelling as it is usually practised means that the modeller regards each flow pathway in its turn and chooses a mathematical statement explaining the flow in terms of the contents of various compartments. That is, explanation at the mesoscopic level is usually achieved by assuming that a flow is the consequence of a causative factor, a 'force'. There is no problem with this procedure if one confines the model to a single process (or possibly a few coupled processes). If, however, one combines many such process models into a large community model, the assumption is usually implicit that the network of flows will be the consequence of the combination of separate forces.

The development hypothesis, however, turns matters around. It says that the development criterion resides in the flow structure and is not the consequence of the aggregate forces. One might say that the system develops in such a way as to select for the proper forces to satisfy the community criterion in the same sense that organisms undergo selection. In such circumstances the results of most community simulations are conditional at best.

This is not to imply that simulation modelling at the community level should cease. Rather it is a caution that it should proceed with an eye upon the eventual development of macroscopic principles that can guide the creation of community models. Optimization theory is developing at a satisfying rate, and one can envision an iterative process whereby a simulation model is produced according to the best estimate of aggregate processes, and the simulation of resultant system behaviour is compared against the macroscopic criterion. The model is then modified until an incremental improvement is made in meeting the criterion. Proceeding in this manner, many of the possible alternate configurations of the system should accrue, allowing for at least a probabilistic forecast of the response of the system to perturbation.

2.2.3 MODELS OF PARTICLE-SIZE SPECTRA[1]

Introduction

Until the middle of the last decade the description of the structure of ecological communities was usually accomplished by grouping organisms according to taxa or trophic level. Compartmental contents and flows between aggregations were described in terms of mass or energy. One of the challenges of experimental community ecology has been the total quantitative description of the contents of and exchanges between compartments. Complete taxonomic identification within a community is difficult, not to mention the herculean task of quantifying all exchanges between the species. The aggregation of species into trophic levels, likewise, has been wrought with ambiguities (Hutchinson, 1959). Any alternative which would allow rapid and repetitive measurement of ecosystem structure would, therefore, hold high potential for

---

1. By R. E. Ulanowicz, Contribution No. 968 of the University of Maryland, Centre for Environmental and Estuarine Studies.

accelerating the progress of ecosystems research by increasing the rate of data acquisition.

A priori there is no reason to aggregate organisms only by taxa or trophic level. Grouping individuals by size, age, or metabolic level might lead to a more accessible interpretation of ecosystem structure and dynamics. In fact, descriptions of particle-size distributions (psd's) in seawater have been made microscopically (Riley, 1963), by filtration (Mullin, 1965), and by light scattering and transmission techniques (Jerlov, 1961). But none of these techniques combined the speed and accuracy inherent in the electrical measurement of the number and volume of particles passing through an orifice. This methodology had proved so useful in biomedical research that by the mid sixties a commercial unit (the Coulter Counter) was on the market, and Sheldon and Parsons (1967) were using the tool to examine psd's in seawater. More recent advances in particle-size measurement, e.g. by means of acoustics and laser holography, have extended the range of particle sizes which may be counted automatically (see Parsons and Seki, 1969, and SCOR, 1973).

In the new perspective afforded by rapid particle counts, plants, animals, detritus, and inorganic material are integrated into compartments characterized by size alone. The key concern of this section is how the structure of particle-size distributions in pelagic ecosystems can assist in the description of community dynamics.

Early empirical work

One of the first problems in applying the new particle-counting technology to ecology was how best to present the data for subsequent analysis. The data are most easily read from the apparatus as the number of particles of a specific volume. Since very small particles tend to dominate in numbers, most graphs of counts versus volume look much the same, having a peak at the low end of the spectrum and quickly decaying to relatively minuscule counts at higher sizes.

But the smaller number of larger particles could occupy a greater proportion of the total sample volume than their numbers would indicate. Multiplying the numbers by the individual particle volume gives the total volume of particles in that size class. Normalizing this product with the total volume of water examined gives the volume concentration of particles of a given size class (usually expressed in ppm). The plot of the volume concentration versus the logarithm (in octaves) of the particle diameter has, therefore, become the standard way of representing particle-size data.

Sheldon and Parsons show how a bloom of the diatom *Skeletonema costatum* (*c*. 40 μ) becomes strikingly apparent when graphed in the above manner, in spite of being invisible on a plot of particle counts versus particle diameter.

With methodological questions settled, the application of particle-size data to ecological research could begin. Parsons (1969) perceived psd's as enabling the rapid determination of the diversity of pelagic ecosystems. By dividing the total particle volume in size class i, by the volume of all particles, one arrives at the probability, p_i, that an atom of biomass will be included in a particle of size i. The community size diversity is thereby described as

$$D = -\sum_i p_i \log p_i, \qquad (66)$$

and may be regarded in the same way as species number or species biomass diversities. The inability, however, of ecological research to establish a firm relationship between diversity and stability caused interest in diversity indices to wane, and Parson's index does not appear to have engendered much subsequent enthusiasm.

Since psd's could be quickly obtained, it became possible to search for a pattern in the large number of size spectra from various marine pelagic communities. Sheldon *et al.* (1972), therefore, collected over a hundred psd's in the $1-100$ μ range from the surface waters of the Atlantic, Pacific, and Southern Oceans and from deep waters of the North Atlantic, South Pacific, and Southern Oceans. Indeed, surface-water spectra did appear to fall into four distinct groups characteristic of the polar, temperate, subtropical, and equatorial regions. The most ubiquitous pattern, however, was no pattern at all, i.e. a uniform distribution of particle concentration over all size classes measured. Uniform particle concentrations dominated the subtropical surface waters and the deep water samples at all latitudes.

Sheldon *et al.* discussed the possibility that uniform concentration distributions might be extrapolated beyond the size ranges studied. Estimates of standing stocks of larger organisms such as zooplankton, micronekton, and whales showed that, while concentrations might fall off slightly for larger-sized organisms, the drop is less than a factor of five over seven orders of magnitude in size. Thus, a uniform distribution over all size classes provides a convenient rule of thumb for estimating the abundance of animals when only their size is known (Sheldon and Kerr, 1972).

Furthermore, there are ecological implications behind the observations of uniform size concentrations. The pelagic food chain relationships in the ocean are generally such that relatively large predators feed on relatively small prey. Therefore, uniform concentrations at all sizes can only be maintained if the rate of particle production varies inversely with particle size. The authors assembled half-life data from organisms spanning six orders of magnitude in size to show that a tight correlation does exist between doubling time and size.

Modelling Sheldon's hypothesis

Any empirical observation, such as Sheldon's sweeping hypothesis of uniform size concentrations, immediately begs for more formal description. How does the empirical fact relate to observations on the size dependency of other physiological factors, such as metabolism? Kerr (1974*a*) attempted a synthesis of observations on psd's, growth rates and metabolic rates by invoking an energy budget after Winberg (1960). Doubling time was assumed to be a power function of organism size

$$\tau = aW^b, \tag{67}$$

where τ is the doubling time, W is the weight (size) of an organism, and a and b are numerical constants. Similarly, metabolism was described by the formula

$$T = \alpha W^\gamma, \tag{68}$$

where T is the metabolic rate and α and γ are numerical constants. Coupling these descriptions with the assumption that predators only devour prey whose weight fraction is a constant fraction of their own, Kerr showed that the ratio of volume concentrations between successive trophic levels varied as the predator weight to the $\gamma + b - 1$ power. Whence, Sheldon's hypothesis of uniform volume concentration requires that $\gamma \approx 1 - b$, that is, the rates of growth and metabolism of various organisms respond similarly, on the average, to changes in body size.

Of course, it remains a logical impossibility that volume concentrations remain constant for arbitrarily large organisms (Bader, 1970). Any real colletion of particles must have a maximum particle size (even the collection of all the organisms in the ocean). Furthermore, Sheldon et al.'s extrapolation to very large sizes showed a slight downward trend, which the authors preferred to dismiss as the underestimation of stocks of large-sized species. Platt and Denman (1977) combined an alternate analysis of energy flow up the food chain with Fenchel's (1974) values for the power law constants in Equation (67) and Equation (68) to show that the volume concentration of particles should decline roughly as the negative one-fifth power of body size. Interestingly enough, Platt and Denman (1978) found that the decline in particle concentrations could be explained almost entirely by losses to metabolic respiration. Losses to the detritus food chain through an assimilated ration have a negligible effect on the predicted psd.

The insight gained by the steady-state analyses prompted attempts to generalize the psd models. With no further assumptions Silvert and Platt (1978) were able to extend Platt and Denman's analysis into a description of transients in a psd. The assumption that losses from a given size category are proportional to the biomass of that compartment guarantees that the first-order partial differential equation describing the psd may be solved analytically. The solution specifies that any transient in the system will propagate up the spectrum without change in shape. If the transient has the form of a pulse disturbance, the relative heights of the peak at two different times will be in the same ratio as the steady-state values. There is no diffusion of the peak as it evolves. One may imagine a snake swallowing a rabbit—the lump of prey can be seen travelling intact down the body, maintaining its shape while diminishing slightly in size during the passage.

Intuitively one would expect pulse disturbance in a psd to spread as it

75

moves to larger sizes. Also, one would expect some 'feedback' of the pulse to smaller-sized particles. That the time-dependent model of Silvert and Platt failed to predict such behaviour prompted the investigators (Silvert and Platt, 1980) to seek a more inclusive form of the time-dependent equation for psd's. The derivation begins with the equations of detailed balance of biomass among the size compartments

$$\frac{db_i}{dt} = S_i = \sum_j (f_{ij} - f_{ji}),$$ (69)

where S_i is the net rate at which material flows into compartment i and is expressed in terms of the transfer rates f_{ij}, which represent flows into compartment i from compartment j. The contributions to the f_{ij} due to reproduction, predation, growth, advection, and diffusion are described in turn, until the master equation takes on a form similar to the equation of species continuity in a fluid flow field—the primary difference being that Silvert and Platt's formulation has particle-size concentration as the dependent variable.

As the master equations form a coupled set of non-linear partial differential equations, a general analytical solution is impossible. Silvert and Platt do, however, solve a limiting form of the master equation for a homogeneous system in which predation (represented in a quadratic fashion) is the dominant biological flow term. The feedback due to predation is in the direction of increasing size, and a step-function increase at small autotrophic sizes is likely to lead to a further increase in small-particle concentration because of reduced grazing pressure. The result is a wave-like instability that may take a long time to die out.

If Silvert and Platt seemed reluctant to forego analytical methods in favour of numerical simulation of the coupled non-linear master equations, Steele and Frost (1977) were not. They derived a model for nutrient-producer–herbivore–carnivore interactions in a pelagic ecosystem. The size of the producers and the length of the herbivores were the only two dependent variables for all processes. Simulations of the model indicated that size structure can be at least as important, and probably more significant, than the total biomass of a population to understanding the exchange of energy between trophic levels. Steele and Frost's model can be viewed as a particular case of the master equations of Silvert and Platt.

Other empirical efforts

It is useful to note that all of the previously described modelling and theoretical efforts can be traced back to the Sheldon hypothesis of uniform volume concentration. This hypothesis, in turn, was based on psd's taken at 100 stations around the world's oceans with fewer than ten distributions from various depths at selected stations. While this is an impressive data base for a single project, it soon becomes apparent that theoretical exercises are extending well beyond their data base. Further development of concepts regarding psd's is

76

likely to be slowed by a dearth of well-planned experiments centred about the measurement of particle size.

Although Sheldon *et al.*'s observation on uniform particle concentrations has received much subsequent attention, it constituted only half of their reported discussions. As mentioned, they also described psd's characteristic of four major oceanic zones. What Sheldon *et al.* did qualitatively can also be accomplished in a more quantitative manner.

For example, Kitchen *et al.* (1975) employed principal-component analysis to describe particle spectra taken off the Oregon coast during the summer. The first two principal vectors accounted for 92 per cent of the intersample variance. When each sample was plotted according to its weighting factors for the two principal vectors, the samples could be segregated into three separate groups according to whether the sample came from the clear deep waters, the high-salinity surface waters, or the low-salinity surface waters.

Chanut *et al.* (1977) performed much the same analysis on spectra from the Gulf of St Lawrence Estuary and identified five separate regions of the estuary. During a subsequent survey, whenever spectra approximating either of two specified psd's occurred, twenty-one other variables characterizing biotic and abiotic materials were also measured. The twenty-one variables were themselves differentiated using principal-component analysis, and the patterns of association of the biochemical parameters with respect to the principal axes differed markedly in the two regions. In the first (upstream) region terrigenous matter associated strongly with the principal characteristic vector, whereas the downstream segregation showed a greater association of biological variables with the primary vector. Thus, the authors were able to characterize the particles according to probable origin.

PSD's and holistic ecology

Having reviewed the theoretical and empirical work on psd's, one might rightfully ask how these endeavours help to define a holistic approach to ecosystems analysis. While it is true that a certain degree of aggregation is accomplished in lumping particles according to size, it is not immediately clear how a particle-size distribution culminates in a single community variable or a small set of variables which characterize the structure of the entire eco-system. It is true that Parsons was tending towards a holistic description of community particle-size structure when he assigned diversity indices to psd's. But particle-size diversity, like its taxonomic counterpart, is a measure based on compartment content and is likely to be of little assistance in elucidating systems dynamics.

Perhaps this last statement requires further explanation. Today it is trite to say that an ecosystem consists of its component species and their interactions. But it is not as widely perceived that if one wishes to describe the behaviour of a community, one should give priority to determining the interactions among compartments over assessing the contents of the compartments themselves.

77

Translated into more practical terms, if mass or energy is the currency in an ecosystem, a primary goal should be the determination of the *flows* of mass or energy—a goal which is at least as important as a description of the mass or energy content of the compartments. Specific examples of relevant flux measurements are rates of excretion, production, respiration, feeding, physical transport and so on. A holistic description of community flow structure appears more likely to be of value in explaining ecosystem development than a description of compartmental contents.

This may indicate why species diversity, a community measure based on content, could not characterize the dynamic notion of stability. In contrast, Mulholland's (1975) rigorous application of information theory to energy flows yields community variables useful in hypothesizing a coherent theory of organization in ecosystems encompassing the notions of self-organization, complexity, redundancy, and stability (Ulanowicz, 1979*b*).

Now particle-size distributions are measures of compartmental content. It is unlikely that a single determination of particle-concentration distribution will say much about the community dynamics of the pelagic ecosystem. In fact the insights provided by the models described in this section actually come about by balancing flows through the chain of compartments—the steady-state psd simply serves as an empirical constraint on the balance. Any effort to generalize the results will invariably lead to what Silvert and Platt aptly term the 'master equations', i.e. the balance of flows among the compartments as described by Equation (69).

Unfortunately, one cannot unambiguously infer the flow structure among size compartments by observing solitary psd's. One can, however, gain considerable information about the flow structure by collecting a time series of psd's at a given location.

Implications for design of research programmes

One is inclined to propose an observational programme along the following lines. Particle-size distributions are measured in a pelagic ecosystem at regular intervals. The length of the interval between determinations would ideally be less than the turnover time of the smallest organisms that will be measured (as estimated from Equation (67)). The duration of the series ideally should be greater than the turnover time of the largest organism in the distribution. In addition to particle size, other convenient variables such as temperature, salinity, and water velocity should be determined at each sampling time. Ideally, one would also hope to measure certain descriptive biochemical parameters such as nutrients, chlorophyll, lipids, and ATP according to the same sampling protocol.

If the particle-counter records over n channels and p additional factors are monitored, one has $n + p$ separate time series. Cross-correlation of the rates of change of the n size channels should yield clues on intercompartmental transfers. Correlation between the particle sizes and the descriptive parameters should provide hints as to the qualitative nature of the flows. Fourier time spectra for each of the n channels may reveal the frequencies in the environment to which each size class is attuned.

Such an endeavour should produce a host of hypotheses concerning possible mesoscopic (intercompartmental) flow mechanisms. What is perhaps more important, it may offer an acceptable method for estimating a measure of the dynamic structure of the community. Assuming that the percentage of the total system variance attributable to channel i is V_i over the period of observation, and calling the partial correlation coefficient of size class i with any other channel j, r_{ij}, then the average mutual information among the n channels becomes

$$I = \sum_{k=1}^{n} \sum_{j=1}^{n} r_{kj}^2 V_k \ln \left[r_{kj}^2 \bigg/ \sum_{i=1}^{n} r_{ij}^2 V_i \right]. \tag{70}$$

This measure is less fundamental than Mulholland's (1975) analogue based on energy flows, but, unlike Parson's diversity index, it is descriptive of the dynamics of the system. Since organisms are lumped according to a controllable number of size classes, there is greater opportunity to compare the indices of two different pelagic ecosystems—regardless of their taxonomic constitution. Furthermore, if one is willing to assume that a greater index of dynamic structure implies increased capability for self-organization, one may order various ecosystems according to their stage of development. Conversely, a decrease in the value of dynamic structure would likely signify a breakdown of the system in the face of inordinate stress.

The automatic measurement of particle-size distributions in marine waters provides several advantages in the analysis of pelagic ecosystems. Unlike taxonomic descriptions, the format of the data is uniform and controllable, i.e. two very different ecosystems may be compared by using the same size-range and number of channels. A virtually synoptic picture of the entire eco-system can be obtained with relative ease and accuracy. The range of sizes over which measurements can be made is expanding (Pugh, 1978).

Whenever possible psd's should be taken along with other data on pelagic ecosystems. Single psd's have little value in elucidating ecosystem dynamics, however, implying that consideration should be given to collecting temporal series of psd's.

2.2.4 STATISTICAL MECHANICS [1]

Introduction

Faced with the apparent complexity and problems of quantification en-countered in ecology, theoreticians have long recognized the desirability of constructing a statistical description of ecological interactions. Thus Lotka

1. By M. J. R. Fasham.

79

(1925) has written: '... what is needed is an analysis ... that shall envisage the units of a biological population as the established statistical mechanics envisage molecules, atoms and electrons; that shall deal with such average effects as population density, population pressure, and the like, after the manner thermodynamics deals with the average effects of gas concentration, gas pressure. ...' The two basic requirements of a statistical mechanics of any system are the existence of a constant of motion and a Liouville theorem which states that the probability of the system being in a certain state in the phase space defined by the population values is constant along a flow line in that space. In physical systems the constant of motion is the energy and the validity of Hamilton's equation of motion ensures the validity of the Liouville theorem (Kittel, 1958). In ecology, however, the problem is more complex in that, so far, no simple equation linking total system energy and population dynamics, in a manner analogous to Hamilton's equation, has been derived (Ulanowicz, 1972; Hirata and Fukao, 1977). We shall see, however, that a constant of motion can be derived for a population governed by the Volterra predator–prey equations. This restriction throws some doubt on the general applicability of the theory and we will return to this point after summarizing the main results. The statistical mechanics of Volterra's equations was derived originally by Kerner (1957, 1959, 1961, 1964, 1972) and developed by Leigh (1965, 1968). A succinct summary has been given by Maynard Smith (1974) while Goel *et al.* (1971) have provided a mathematical review and also extended the theory.

The basic theory

Consider a set of n species having populations N_1, N_2 ... N_n, then Volterra's equations can be written

$$\frac{dN_i}{dt} = \varepsilon_i N_i + \sum_{s=1}^{n} \alpha_{is} N_i N_s; \qquad i = 1 \dots n. \tag{71}$$

The ε_i term represents how the species would develop in the absence of the interaction terms: it is a positive quantity for prey species and negative for predators. The α_{is} term represents the strength of the interaction between species i and s and in Volterra's equations these terms are antisymmetric, i.e.

$$\alpha_{is} = -\alpha_{si}. \tag{72}$$

This antisymmetry means that the species numbers will display a purely oscillatory nature if displaced from equilibrium. As pointed out by Maynard Smith (1974) this implies that if two samples were taken from the same ecosystem at two different times, such that half a population cycle had elapsed, the common species in the first sample would be rare in the second sample and vice versa. Another consequence of antisymmetry is that $\alpha_{ii} = 0$ and thus there are no self-limitation terms (often called Verlhurst-Pearl terms). If such terms were introduced, they would damp out the oscillations of the species in a way which probably accords better with the ecologist's intuitive feeling of how an ecosystem behaves. However, the introduction of damping terms means that no statistical mechanics can be developed (Goel *et al.*, 1971).

Returning to the Volterra equations, the equilibrium population levels q_s are given by the equation

$$\varepsilon_i + \sum_{s=1}^{n} \alpha_{is} q_s = 0. \tag{73}$$

If we now define a new variable $x_i = \log(N_i/q_i)$ then Equation (71) can be rewritten

$$\frac{dx_i}{dt} = \sum_{s=1}^{n} \alpha_{is} q_s (e^{x_s} - 1). \tag{74}$$

Equation (74) can now be integrated to show that

$$G = \sum_{i=1}^{n} q_i (e^{x_i} - x_i) = \text{constant}. \tag{75}$$

The quantity G is thus the constant of motion for the Volterra dynamics and its existence is a direct consequence of the antisymmetry of the α terms. Equation (74) can now be rewritten in the form

$$\frac{dx_i}{dt} = \sum_{s} \alpha_{is} \frac{\partial G}{\partial x_s}. \tag{76}$$

The next step in the development of a statistical mechanical theory is the derivation of a Liouville theorem. Consider a large number of possible biological associations each developing according to Equation (76) but having different starting conditions. We can define a density function $\rho(x_1, x_2 \ldots x_n)$ in the phase space defined by $x_1, x_2 \ldots x_n$, which measures the frequency that the system attains the state $x_1, x_2 \ldots x_n$. Using a fluid dynamical analogy we can state that the function must obey the equation of continuity, namely

$$\frac{\partial \rho}{\partial t} + \text{div}(\rho \underline{V}) = 0 \tag{77}$$

where $\underline{V} = (\dot{x}_1, \dot{x}_2 \ldots \dot{x}_n)$. Expanding the divergence operation and using Equation (76) we find that

$$\frac{\partial \rho}{\partial t} + \sum_{i} \dot{x}_i \frac{\partial \rho}{\partial x_i} = 0. \tag{78}$$

This is Liouville's theorem, which states that as we move along with the system the density in our neighbourhood remains unchanged. The results of Equations (75) and (78) enable the statistical mechanical theory of Gibbs ensembles to be applied to ecological systems.

Canonical ensembles

In classical statistical mechanics the canonical ensemble is regarded as a subsystem which can exchange energy with the surrounding system whose total energy content is constant. Using these assumptions it can be shown that the probability density of the subsystem having energy E is given by the familiar Boltzmann formula,

$$\rho(E) = A \exp(-E/kT), \tag{79}$$

where A is a constant and T the absolute temperature. Applying the same reasoning to the Volterra equations, the probability that the system will have a given G value will be

$$\rho(G) = C \exp(-G/\theta) \tag{80}$$

where θ is the ecological temperature which will be defined later. Now $G = \Sigma\, G_i$ where $G_i = q_i(e^{x_i} - x_i)$, and so the distribution of the individual species values is independent. Transforming back to the N_i values we find that the probability that the ith species has the value N_i can be written

$$P(N_i) = \frac{1}{\Gamma(q_i/\theta)} \theta^{-q_i/\theta} N_i^{q_i/\theta-1} e^{-N_i/\theta}. \tag{81}$$

This is a Pearson type III distribution and Kerner (1957) notes that Fisher et al. (1943) used this distribution to derive the well-known logarithmic law for species abundance distributions which has some observational support. However, in view of the difficulty experienced in using observed species abundance data critically to differentiate between different theoretical models of species abundance curves (Bliss, 1965), it is doubtful if this result can be used to provide observational support for Volterra statistical mechanics. Furthermore, as will be shown later, the distribution given by Equation (81) can be derived from other assumptions.

Using Equation (80) we can now calculate the canonical averages of a number of variables. The canonical average $\overline{D}_i$ of

$$D_i = \frac{\partial G}{\partial x_i} \equiv q_i\left(\frac{N_i}{q_i} - 1\right) \tag{82}$$

is zero. Thus, as might be expected, the canonical average of N_i is q_i, the equilibrium value. The ecological interpretation of θ is best achieved by deriving the canonical average of D_i^2 which is

$$\overline{D_i^2} = \theta q_i. \tag{83}$$

Rearranging and substituting for $\overline{D_i^2}$ in terms of N_i and q_i we find that

$$\theta = \frac{1}{q_i}\overline{(N_i - q_i)^2}. \tag{84}$$

Thus to quote from Kerner (1957), '... the temperature measures, in one number common to all species, the mean square deviations of the populations from their stationary value q_i, and vice versa. Zero temperature corresponds to the completely "quiet" stationary state of biological association. The temperature is, so to speak, a kind of indicator of the level of excitation of the association from its stationary state.' Once θ has been determined, the interaction terms can be estimated from the canonical average of $x_i \dot{x}_r$ which is given by

$$\overline{x_i \dot{x}_r} = \alpha_{ri} \theta. \tag{85}$$

A parameter of the population dynamics that would be useful experimentally is the frequency with which the variable $x_i(t)$ crosses a given line $x_i = a$. Kerner (1959) showed that this frequency W_a is given by the time average

$$W_a = \lim_{T \to \infty} \frac{1}{2T} \int_{-T}^{+T} \delta[x_i(t) - a] |\dot{x}_i| dt, \tag{86}$$

where δ is the Dirac delta function. In order to derive W_a from the statistical mechanics we have to make the Ergodic assumption that time averages of a function are the same as ensemble averages. The conditions under which this assumption is justified are discussed fully by Goel $et\ al.$ (1971), where it is shown to be true as long as the number of species is large. The required ensemble average is given by

$$W_a = C \int \delta[x_i - a] |\dot{x}_i| \exp(-G/\theta) dx_1 dx_2 \ldots dx_n. \tag{87}$$

The evaluation of this integral presents mathematical problems but it is possible to calculate the frequency W_a relative to, say, the frequency W_o and this is given by

$$W_{rel}(a) = e^{q_i/\theta} (ae^{-a})^{q_i/\theta}. \tag{88}$$

Bearing in mind that $x_i = \log(N_i/q_i)$, the quantity $W_{rel}(a)$ gives the relative frequency that a species crosses the line $N_i = q_i e^a$ relative to the line $N_i = q_i$, the equilibrium or average value. It is evident from Equation (88) that N_i will cross the average line more frequently than any other line and that the relative frequency falls off quite rapidly with increasing a.

Leigh (1965, 1968) used a linear approximation to the Volterra equations to derive an ensemble average of W_a given by

$$W_a = \frac{1}{\pi} \exp\left[\frac{-a^2}{2q_i}\right] \sqrt{\sum_r q_i q_r \alpha_{ir}^2}. \tag{89}$$

Leigh then asks under what conditions will W_a be minimized and thus the persistence of the ecosystem be maximized. He defines the productivity P as

$$P = \frac{1}{2} \sum_{rs} |\alpha_{rs}| q_r q_s \tag{90}$$

and minimized the function $\sum_{r=1}^{n} q_i q_r \alpha_{ir}^2$ given the restraint of constant P.

This occurs when the α_{ir} are all equal and leads to the conclusion that persistence is increased if the food-web consists of a large number of small interactions rather than a small number of strong ones. If we assume that the $|\alpha_{ir}|$ are all equal then Leigh (1965) shows that for a diverse community

$$W_a = \alpha \frac{2}{\sqrt{n}} \frac{P}{B} \exp(-a^2/2q_i\theta), \qquad (91)$$

where B is the total biomass defined as $\Sigma_i q_i$. This shows that the frequency with which a species in a population crosses the line $x_i = a$ increases with increasing productivity and decreases with increasing biomass. This is an interesting result but it should be remembered that it depends on a linearity assumption and so is only definitely true of a system close to equilibrium.

Relationship to stochastic differential equations

In the previous section we have shown that for a canonical ensemble of species obeying Equations (76), the distribution of the quantity $x_i = \log(N_i/q_i)$ is given by $C\exp(-G_i/\theta)$. Leigh (1968) showed that a similar distribution for x_i is obtained if the $x_i(t)$ satisfy the stochastic differential equation

$$dx_i(t, a) = -b\partial G_i/\partial x_i \cdot dt + dB(t, a) \qquad (92)$$

where a is an index of the possible paths $x(t)$, varying between 0 and 1 and $dB(t, a)$ is a Gaussian Markov input. In turn Equation (91) can be shown to be equivalent to a logistically growing population subject to environmental shocks. We thus have the disquieting situation that two different models, one based on food-web relationships and one based on environmental stochasticity, both give the same probability distribution for the population numbers. Leigh (1968) showed that the two models can be distinguished theoretically by the first derivative of the autocorrelation function $\mu(s)$ of $x_i(t)$ at $s = 0$, but whether this fact could be used in practice remains uncertain.

This connection between the statistical mechanics and stochastic differential equations perhaps also suggests that future progress in statistical population dynamics might come from the direct investigation of more general stochastic differential equations rather than via statistical mechanics with its more restrictive assumptions. Leigh (1968) has made a beginning by studying a generalization of the linear form of (92) which allows for self-crowding effects.

Experimental tests of the theory

Goel et al. (1971) have reviewed the data that provide some experimental support for the validity of Volterra's equations. In his second paper on the topic Kerner (1959) applied the statistical mechanics to data on catches of foxes in Labrador from 1834 to 1925. He calculated the time averages of a number of variables derived from the data. Using the Ergodic assumption he was then able to use the results on canonical averages to calculate a series of estimates of the ecological temperature θ. All these estimates were in reasonable agreement, and so he used the value of θ to calculate W_a for a range of values of a and obtained a very good agreement between theory and observation. This appears to be the only attempt to apply the Volterra statistical mechanics to observational data. Leigh (1968) analysed the well-known lynx and hare data but did so using a more general linear stochastic differential equation rather than statistical mechanics.

Kerner (1957) suggested using the equipartition result (Equation (84)) as a test of the theory. Thus, if a time series of population values for a large number of species in an ecosystem were available then, assuming ergodicity, the time average of $(N_i/q_i - 1)^2$ should be the same for all species. (In this context it is interesting to note that the lynx and hare data yield values of θ of 1·51 and 1·54 respectively.) The main problem is that there are very few data sets providing observations on a number of species over a large number of population cycles. One possible candidate is the plankton data recorded since 1948 by the Edinburgh laboratory of the Institute of Marine Environmental Research, although account would have to be taken of the obvious secular change of q_i displayed by some species (Colebrook, 1972).

General assessment of the theory

The usefulness of the statistical mechanics of Volterra's equations has been strongly criticized by May (1973) who states: '(The) results are undeniably very elegant, but they are fragile, as they ultimately rest on the precise anti-symmetry of the α_{ij} coefficients. The underlying conservation law is much less robust than the spatial and temporal invariance of the laws of physics which underpin statistical mechanics. . . .' He goes on to say that '. . . I am sceptical of any interpretation of the fluctuations observed in natural populations which is based on the pathological neutral stability character of a set of specially antisymmetric Lotka-Volterra equations'. However, as pointed out by Goel *et al.* (1971), if similar criticisms had discouraged physicists investigating models used in many-body physics, the development of this topic would have been severely hindered.

In the last analysis, therefore, the theory must stand or fall on whether it can explain observed data. In principle this empirical verification can be done and is the first priority in any development of the topic. In the meantime, theoretical work might be better concentrated on the statistical properties of solutions of more general stochastic differential equations (see the next section). Alternatively it might be feasible to develop a statistical mechanical theory based on less restrictive assumptions than the Volterra equations. Demetrius (1977) has proposed a model in which interactions between species are described in terms of energy flow. He then applies the statistical mechanical theory of lattice systems to derive certain macroscopic parameters. However the derivation is couched in rather obscure mathematics and so it may be some time before the applicability of this development can be assessed.

2.2.5 STOCHASTIC MODELLING AND POWER SPECTRAL REPRESENTATION[1]

Introduction

Much of the recent development of theoretical ecology has concentrated on deterministic models (e.g. May, 1976), whereas it is common experience that ecological processes are subjected to considerable stochastic variability. The statistician's answer to this problem would be to estimate some statistics of the ecological variable being studied. In some cases it may be possible to calculate a first-order statistic, such as a running mean, and attempt to predict the behaviour of this statistic by using a deterministic model. In other cases this may not be possible, in which case some second-order statistic of the process such as the covariance or spectral function might be calculated. The ecological variable being studied will in general be a function of time and space $N(\underline{r}, t)$ and the covariance function defined as

$$R(\underline{r}, \underline{s}, t, u) = \xi\{[N(\underline{r}, t) - \xi\{N(\underline{r}, t)\}][N(\underline{s}, u) - \xi\{N(\underline{s}, u)\}]\}. \quad (93)$$

If the processes are stationary in space and time then the covariance will depend only on the differences $\underline{x} = \underline{r} - \underline{s}$ and $\tau = t - u$ and we can define a spectral function,

$$F(\underline{k}, w) = \int_{-\infty}^{+\infty} R(\underline{x}, \tau) \exp[-i(\underline{k} \cdot \underline{x} + w\tau)] d\underline{x} d\tau \quad (94)$$

which is analogous to the familiar one-dimensional spectral function. Another possibility is that the process might be stationary in space but not time and vice versa.

Having calculated a second-order statistic there are two possible ways to proceed. The first would be to attempt to explain the form of the statistic using some rather general model such as the autoregressive or moving-average models (e.g. Jenkins and Watts, 1968; Poole, 1976). This approach has, for example, been applied to the well-known time series data on the hare–lynx cycle (Bulmer, 1974; Campbell and Walker, 1977). The problem with this approach is that having done this, one still has to interpret the results in ecological terms. The second and perhaps more fruitful approach is to develop stochastic models which explicitly incorporate what we know or suspect about the actual ecological interactions. If these models can be used to predict the behaviour of our measured statistics, these predictions can then be compared with observations to validate or reject the model.

In the subsequent discussion we will, for the sake of simplicity, consider a system consisting of, at most, two species which can vary in time and, perhaps, one space dimension. Let the function $(N_1(x, t), N_2(x, t))$ represent the abundance of the two species, then a reasonably general model can be written as

---

1. By M. J. R. Fasham.

$$\frac{\partial N_i}{\partial t} = F_i\left(N_1, N_2; \frac{\partial N_1}{\partial x}, \frac{\partial N_2}{\partial x}; \frac{\partial^2 N_1}{\partial x^2}, \frac{\partial^2 N_2}{\partial x^2}\right), \quad i = 1, 2 \qquad (95)$$

where the function F_i will contain both deterministic and stochastic elements. This model is often further simplified to

$$\frac{\partial N_i}{\partial t} = E_i(N_1, N_2) + T_i\left(\frac{\partial N_i}{\partial x}, \frac{\partial N_2}{\partial x}, \frac{\partial^2 N_1}{\partial x^2}, \frac{\partial^2 N_2}{\partial x^2}\right), \quad i = 1, 2. \qquad (96)$$

Here the function F_i has been split into a function E_i which describes the ecological interactions between the species (i.e. growth, predation, competition, etc.) and a function T_i describing the spatial transport of the species. It should be emphasized that this simplification precludes a number of possible processes such as density-dependent dispersal (McMurtrie, 1978) or cases where the functions E_i are explicit functions of x.

The literature of stochastic differential or difference equations is extensive and complex (see for example Bartlett, 1966; Soong, 1973; Goel and Richter-Dyn, 1974), and we shall restrict the discussion to models for which covariance or spectral functions have been calculated.

We may distinguish the following classes of models:
1. Spatially homogeneous models. These can be further subdivided into those that are stationary or non-stationary in time.
2. Spatially heterogeneous models that are stationary with respect to space but may or may not be stationary in time.
3. Spatially heterogeneous models that are not stationary with respect to space.

For all of these classes of models the stochastic variability can be introduced as either extrinsic or intrinsic processes. In the first case the stochastic element of the equation will be a simple forcing function, while in the second case one, or more, of the coefficients describing the ecological processes will be a stochastic function. Exactly which approach is adopted will depend on the situation being modelled.

Spatially homogeneous models

In these models it is assumed that the variables N_1, N_2 do not vary greatly with distance and so the transport function T_i can be ignored and the equations reduce to ordinary differential equations. Because of this simplification, spatially homogeneous or 'lumped parameter' models have received the most attention from theoreticians. It should always be borne in mind, however, that there is no point in studying these simpler formulations if the real 'ecological world', which is by observation spatially heterogeneous, is misrepresented by them. For instance it is well known that diffusion coupled with non-linear species interactions can give rise to new and unusual phenomena often called diffusive instabilities (Okubo, 1974; Nicolis and Prigogine, 1977) which have no counterpart in the spatially homogeneous world. If these or similar phenomena turn out to be significant in ecological situations, then the study of

87

spatially homogeneous models may prove a blind alley. Conway *et al.* (1978) have provided a mathematical stimulus to this problem by determining the conditions under which the 'lumped parameter' assumption can be justified for deterministic parabolic equations.

The simplest form that can be assumed for the function E_i is a linear one given by

$$E_i(N_1, N_2) = a_{i1}N_1(t) + a_{i2}N_2(t) + Y_i(t), \tag{97}$$

where a_{i1}, a_{i2} are constants and $Y_i(t)$ is a stochastic forcing function. If the statistical properties of $Y_i(t)$ are known then standard techniques (Bartlett, 1966; Jenkins and Watts, 1968) can be used to calculate the covariance function and (assuming the $Y_i(t)$ are stationary) the spectral function. In most ecological situations a linear relationship between the species cannot be assumed, and so the equations must be linearized in some way. If the deterministic functional relationship between the N_i is given by $H_i(N_1, N_2)$ then this linearization is usually achieved by writing

$$a_{ij} = \left[\frac{\partial H_i}{\partial N_j}\right](N_1 = N_1^*; N_2 = N_2^*), \tag{98}$$

where the differential is evaluated at the equilibrium values N_1^*, N_2^* and it is understood that the equation now models deviations from these equilibrium values. Obviously this restricts the applicability.

Bulmer (1976) has attempted to use this model to study predator–prey oscillations. He took the variables $N_i(t)$ to represent the logarithm of species abundance and, assuming that $Y_1(t)$, $Y_2(t)$ was a bivariate white noise process (i.e. uniform amplitude over all frequencies), he calculated the spectral function for N_1 and N_2. He also calculated spectra for the case where the two species are undergoing limit cycle behaviour and attempted to use this result to model the hare–lynx cycle referred to earlier. He found that the observed spectra for these data could not be explained by a hare–lynx limit cycle. He suggested instead that oscillations derive from a hare–plant limit cycle, which then drove the lynx cycle via a simple linear model of the form of Equation (97) and obtained reasonable agreement with the data. Leigh (1968) also attempted to fit the hare–lynx cycle using a linear stochastic model, but with little success.

Cumberland and Rohde (1977) consider a stochastic equation where E_i takes the form

$$E_i(N_i) = N_i(t) Y_i(t), \tag{99}$$

where $(Y_1(t), Y_2(t))$ is assumed to be a bivariate Ornstein-Uhlenbeck process. Thus in this model there is no specific functional relationship between the two species and any dependence between them must be modelled by inter-dependence between Y_1 and Y_2. This is rather a restriction, but the authors give a theoretical example of how the predator–prey situation might be described by this model. The advantage of the model is that it is possible to determine analytically the mean and covariance function of the derived variable

$Z_i(t) = \log(N_i(t)/N_i(o))$. The authors also show how the parameters of the model might be estimated from observations.

Roughgarden (1975a) has studied a stochastic version of the Lotka-Volterra competition equations in discrete time given by the difference equation

$$N_{i,t+1} = \left[r_i + 1 - \sum_{j=1}^{s} \frac{r_i \alpha_{ij}}{K_{i,t}} N_{j,t} \right] N_{i,t}, \tag{100}$$

where r_i are the intrinsic growth rates, α_{ij} the competition coefficients, $K_{i,t}$ carrying capacities and $\underline{s}$ the number of species. He assumes that any stochastic variability affects only the parameter $K_{i,t}$ and by linearizing the equation he derives a formula for the spectrum of the N_i as a function of the spectrum of the K_i. As a particular example he considers the bivariate case where $K_1(t)$, $K_2(t)$ are white-noise processes that are either uncorrelated, or positively or negatively correlated. However, there is no attempt to apply the technique to any data. A single-species version of this model has also been described (Roughgarden, 1975b).

Spatially heterogeneous models stationary in space

If the process being studied is stationary in time and space then we can in principle calculate for N_1, N_2 the vector spectral function $\underline{F}(k, w)$. There do not appear to be any bivariate models for which this has been done, although Whittle (1962) calculated the frequency-wavenumber spectrum for the simple univariate Malthusian growth model

$$\frac{\partial N}{\partial t} = N + D\frac{\partial^2 N}{\partial x^2} + Y(x, t) \tag{101}$$

where $Y(x, t)$ is a stochastic forcing function. Whittle was mainly interested in the form of the wavenumber spectrum when $t \to \infty$ and so he does not apply his two-dimensional spectrum to any data. Experience obtained from the study of internal waves (Garrett and Munk, 1975) would suggest that, if suitable data sets were available, an increase in understanding might result from the study of frequency-wavenumber spectra.

If, as is often the case in studies of plankton patchiness, we are mainly interested in spatial spectra, and if it is possible to observe the spatial spectrum in a time scale that is short compared to the time scales of the biological field, then there may be an advantage in studying how this spatial spectrum changes with time. Analytically this can be done by considering the function

$$f_{ij}(k, t) = E[Z_i(k, t)Z_j^*(k, t)] \tag{102}$$

where
$$Z_i(k, t) = \int_{-\infty}^{+\infty} N_i(x, t)e^{-ikx}dx. \tag{103}$$

The advantages of this formulation are that we can also consider the processes that are non-stationary with respect to time (e.g. systems subject to transient effects). Furthermore, any system that can be described by a linear partial differential can be reduced to an ordinary differential equation. Fasham (1978a) has derived the function $f(k, t)$ for the Whittle model (Equation (102)) and has reported simulations of a non-linear model where the Malthusian growth term is replaced by the logistic function (Fasham, 1978b). These studies suggest that the shape of the spatial spectrum of phytoplankton will change significantly during the progressions of a spring bloom, and there is some experimental observation to support these results (Steele and Henderson, 1978).

Fasham (unpublished) has recently calculated the spectral function for the bivariate generalization of Whittle's model (Bartlett, 1975) given by

$$\frac{\partial N_i}{\partial t} = a_{11}N_1 + a_{12}N_2 + D_1\frac{\partial^2 N_1}{\partial x^2} + \phi(x, t) \tag{104}$$

and

$$\frac{\partial N_2}{\partial t} = a_{21}N_1 + a_{22}N_2 + D_2\frac{\partial^2 N_2}{\partial x^2} + \psi(x, t)$$

where the $\phi(x, t)$, $\psi(x, t)$ are random forcing functions with a white-noise spectrum in time but not necessarily in space. The resulting spectral function is rich in structure and it is too early yet to gauge the usefulness of this model. However, it may provide a means of predicting the development of the spatial spectra of, say, chlorophyll fluorescence and nutrients during a spring bloom which could be tested by realizable experiments.

A related model describing the dispersion of a single species population in discrete space-time has been described by Roughgarden (1977).

Spatially heterogeneous models non-stationary in space

This situation would occur if the species interaction, transport parameters or forcing functions were non-stationary functions of the space parameter. This effect is often observed experimentally, and in these cases the non-stationary component is usually removed by trend analysis or filtering before calculating the spectra. It is then often assumed that the resulting spectra can be modelled using stationary processes. Whether this is justified or not will depend on the precise forms assumed for the spatial dependence. As far as we are aware models of this sort have not been studied analytically.

Discussion

It will be apparent from the preceding discussion that the number of stochastic ecological models for which spectral or covariance functions have been calculated is very few and their applications to real data even fewer. It is to be hoped that theoretical ecologists will develop these models in future, as they would seem to offer a number of advantages. The first of these is in model validation. Second-order statistics, such as the spectral function, provide a summary of the data from which a lot of extraneous noise, present in the raw data, has been removed, and these functions may be easier to model than the raw data (Platt and Denman, 1975). This is especially true in the study of the evolution of a population in space-time. We have discussed some analytical methods of modelling spectra but there is also the possibility of using simulation techniques for non-linear cases that are analytically intractable (Steele and Henderson, 1977; Fasham, 1978*b*).

A second benefit may accrue if it is found that a lot of experimental data can be modelled by spectral or covariance functions which are specified by a small number of constants. If these constants can be accurately estimated from the data then they may provide a holistic description of species interactions.

A number of problems with the stochastic approach are apparent, the main one being the difficulty of calculating second-order statistics for non-linear models. Apart from simulation techniques, the use of moment equation methods (Soong, 1973; Bover, 1978) may hold some possibilities. Another problem facing marine ecologists is the relationship between the simple models described above and models based on the statistics of fluid turbulence (Denman *et al.*, 1977).

3

Implications for design of research programmes

3.1   Physiological rates and ecological fluxes

3.1.1   NECESSITY OF MEASURING FLUXES[1]

It is a conclusion of almost every section in Chapters 1 and 2 that the greatest single lack in biological oceanography is knowledge of the fluxes in marine ecosystems. The usefulness of existing simulation models is limited by the fact that we often have good data on the biomasses which constitute the state variables within the compartments, but we lack good data on the fluxes between them. Consideration of the holistic properties of systems also enables us to see that knowledge of fluxes is of the greatest importance. For those accustomed to using the organism as their conceptual unit, there is a considerable effort involved in shifting attention to fluxes.

The most common fluxes in a biological system are the trophic transfers (e.g. predation, grazing, nutrient uptake, etc.) together with egestion, excretion, and detritus formation. All removals from a compartment must in the long term be balanced by production, i.e. growth and reproduction, or by advection, if that compartment is to persist. During the International Biological Programme (especially those parts of it carried out outside the United States) one of the major themes was the measurement of productivity in a variety of ecosystems. This was the beginning of a general awareness of the importance of a dynamic approach. During that same programme, however, scant attention was paid to the flux of nutrients, although numerous measurements were made of nutrient concentrations (Brylinsky and Mann, 1973). It is now clear that a deeper understanding of ecosystem function depends on good data being obtained on the fluxes of organic compounds in particulate and dissolved form, and on the excretion, transport and uptake of inorganic substances. The next section deals with some techniques that have been found useful for helping to evaluate physiological rates and ecological fluxes.

1. By K. H. Mann.

3.1.2 FEEDING AND FOOD-WEBS[1]

Introduction

Compartmental models of ecosystems may differ markedly in all respects except one. All such models incorporate intercompartmental fluxes of mass or energy and the system description requires estimates of these fluxes.

The task of estimating these intercompartmental transfer rates is made difficult, if not impossible, by the problem of first determining which among the compartments do exchange material or energy. This task of determining the topology of intercompartmental connections is laid at the door of the marine biologist under the innocuous title of food-web analysis. The problem of ascertaining 'who eats whom . . .' is not an easy one. Classical approaches such as gut-content analysis are tedious, heavily biased by differential digestion rates, extremely labour-intensive, and not really pleasant undertakings. The difficulty is compounded by the enormous number of intercompartmental connections which must be specified for even relatively small ecosystem models. The problem of food-web analysis is nowhere more acute than in the study of marine ecosystems, where the total of all the man hours of observation is still minimal.

In this section we have attempted to refer to those currently employed techniques by which consumer and food-resource relationships can be most rapidly determined. Owing to the complexity of marine food-webs, we have emphasized techniques which are capable of experimental verification, omitting references to such approaches as examination of mouth parts, which demand subjective judgements. Although both laboratory and *in situ* experiments are described, the reliability and intrinsic value of the latter over the former cannot be too strongly emphasized.

The techniques of food-web analysis which we consider are observation of *in situ* feeding, laboratory feeding experiments and *in situ* caging or exclusion experiments, gut-content analysis, and the employment of tracers.

Wherever possible, techniques amenable to field studies have been stressed and current technological innovations which may facilitate such field studies have been referenced.

Observation of feeding events

The most straightforward approach to determining the food resources available to an organism would seem to be observing that organism during feeding. This direct approach suffers from several inherent restrictions, which also make it of little use in studies to estimate rates of feeding. Such observations require either relatively long periods of submarine monitoring by expensive equipment and subsequent analysis of the recorded observations, or a large number of hours spent under water by trained biologists. In either case, the results obtained

1. By D. F. Smith.

are extremely expensive in terms of man hours required, and the rate at which results are obtained is extremely slow. A second difficulty with this approach is not so immediately apparent: having observed a feeding event it may still not be possible unambiguously to determine which component of the ingested material is the actual food resource. As a case in point, it is now believed that many 'detritivores' are actually feeding on the microbial flora attached to detrital particles, as the nutritional value of the former is large while that of the latter is often vanishingly small.

These difficulties notwithstanding, actual observation of feeding remains the most direct analysis of a simple food-web and can afford information impossible to obtain by other means. Much of our recent knowledge concerning the feeding habits of appendicularians (Alldredge, 1976) and other soft-bodied, gelatinous zooplankton (Hamner et al., 1975) was obtained by direct observations using SCUBA and resulted in information which could never be obtained by classical sampling methods. The use of SCUBA for nocturnal observations of the Pacific electric ray (Bray and Hixon, 1978) resulted in the discovery that this organism, known to be sluggish by day, is at night an active predator. Jennings and Gibson (1969) reported the feeding habits of seven species of rhynchocoelan worms after observing the organisms and comparing the composition of the prey and the enzymatic secretions of the predators subsequent to feeding.

Observations of marine organisms feeding in situ are necessarily limited in scope by the duration and depths which a diver can endure in safety while still functioning efficiently. Attempts to extend the duration and permissible areas of direct observation are being made in several parallel ways. One of these is the development of the submersible (Piccard, 1966; Pritzlaff, 1970) which increases maximum attainable depth and gives greater capacity for carrying instrumentation.

A second method of extending direct observation of the marine environment has arisen with the development of the underwater habitats (Kahn and Haynes, 1970; Collette and Earle, 1972) which permit observers to spend weeks, instead of hours, under water.

A third method by which underwater observation of marine communities has been extended arose with the development of underwater television systems (Myrberg, 1973a). Underwater television and video recording have been successfully employed to study the feeding behaviour of fish (Cummings et al., 1966; Stevenson, 1972), the ethology of pomacentrids (Myrberg, 1972; Myrberg, 1973b; Myrberg and Thresher, 1974) and in commercial fisheries research (Sand, 1956).

Feeding experiments

Among alternative approaches to food-web analysis which yield quantitative rate estimates if short-term incubation is used are a variety of feeding experiments which employ one of two types of controls. Either the consumer(s) have been removed from the control site (exclusion experiments) or the consumer(s) and food resource(s) are incubated together. These two types of experiments may be conducted *in situ*, making the results more meaningful, but the techniques are more laborious and time-consuming than laboratory experiments. The more frequently employed laboratory experiments employ a single food resource, food resources supplied as mixtures of known composition, or the naturally occurring assembly of food resources available to the test organisms.

Exclusion (caging) experiments have been conducted in a wide variety of marine environments (Stephenson and Searles, 1960; Jones and Kain, 1967; Dayton, 1971). The approaches which they have used are directly applicable to food-web analyses if the species being examined as consumers can be isolated from the study site. Such experiments are conducted by isolating a series of replicate sites in cages and removing the test species from the experimental cages. At the termination of the experiment, evidence of feeding is taken to be an observed increase in numerical density or growth of the food resource in the experimental versus the control cages. Such an approach was utilized by Young *et al.* (1976) to separate the predatory impact of fishes from that of the decapods upon the macrobenthos associated with seagrass. Cages were erected at three sites and replicate core samples removed at four different times during the five-month study. The thirty-three numerically most abundant taxa were selected for testing of differences in species densities by two-way multivariate analyses of variance with interaction. Feeding and the overlap of grazer diets were examined by Nicotri (1977) by adding grazers to diatom assemblages growing on artificial substrates. In this instance isolation was achieved by the intertidal mud which proved to be an effective barrier to both immigration and emigration of the test species.

Relative rates of ingestion of the constituents of a mixture of food resources can be employed to estimate the degree to which a species is linked to a food-web, relative to other species. Frankenberg and Smith (1967) tested a polychaete, a gastropod, a pelecypod, two species of crustacea and a teleost as consumers of the faecal pellets of thirty different species. In analogous experiments Mullin (1966), Mullin and Brooks (1967), and Rieper (1978) studied the relative rates of consumption of various diatoms and *Artemia salina* nauplii, and Rieper (1978) studied the relative incorporation rates of different marine bacteria presented to test organisms. The relative rates of ingestion were determined by measuring the food-particle densities before and after the experimental incubations.

Particle counting and particle-size spectra, measured before and after incubation with test organisms, have been employed in experiments in which the naturally occurring particles of seawater are supplied as the food resource. This technique has been employed by Poulet (1976) to study the feeding of *Pseudocalanus minutus* on non-living and living particles from nature and from cultures. Particle-size distribution of suspended matter in seawater was measured before and after incubation with *P. minutus* to ascertain which particle

95

sizes were being ingested. The appropriate particle size fractions were then analysed for carbon, nitrogen and ATP content, to differentiate living from non-living material. Selective feeding of *Calanus helgolandicus* was studied by Paffenhöfer and Strickland (1970) by incubating this copepod with living phytoplankton, natural detritus, or detritus prepared from phytoplankton cultures. The consumption of food was estimated from the changes in particulate carbon concentration as determined by the persulfate oxidation technique (Strickland and Parsons, 1972).

An interesting variation in laboratory feeding experiments might be possible by modifying the recently published technique of Poulet and Marsot (1978). These authors discovered the chemosensory role played in copepod grazing by supplying spherical microcapsules, with and without impregnated nutrient, to mixed populations of copepods. The feeding activity of the copepods was estimated by comparing the change in particle size distribution in the experimental containers relative to that of the control chambers. For those organisms relying heavily on chemosensory perception for feeding, this approach might prove to be a convenient laboratory tool in selective feeding experiments. The feeding experiments described above, if conducted over small time scales, can furnish quantitative estimates of the feeding rates required by ecosystem models.

Gut-content analyses

Examination of the gut contents of an animal was one of the earliest methods attempting to determine predator–prey relationships. This approach is still widely employed, and has the merit that identifiable stomach contents are undoubtedly the remains of prey. The most serious drawbacks to employing gut content analyses are: (a) only those prey species with parts sufficiently resistant to digestion (and frequently, to preservation) will be identified as a food resource; (b) the technique is applicable only to organisms above a minimum size class, for stomach analysis of smaller forms becomes progressively more difficult; (c) the technique is inherently labour-intensive. In spite of these difficulties much information that would have been difficult, if not impossible, to acquire in any other manner has been obtained with this technique. The size distribution of an appendicularian, which is fed upon by pelagic plaice larvae, has been estimated from the size of the faecal pellets found in the larval gut (Shelbourne, 1962). The times of feeding of plaice and sand-eel larvae were determined by gut-content analyses of samples taken at 2-hour intervals by Ryland (1964). On a longer time scale, Bregnballe (1961) studied the annual variation in larval flounder diet over a two-year period. The lack of dietary overlap and therefore absence of competition from four species of Pleuronectioform larvae was reported by Last (1978) from analysis made of the stomach contents of the larval fish. Only organisms found in the stomach were considered since those in the lower gut were too badly deteriorated to permit identification. The combination of results reported by Legand *et al.*

(1971) and by Roger (1975) allowed Roger and Grandperrin (1976) to describe an entire food-web extending from phytoplankton to tuna, solely on results obtained by gut-content analyses. Roger (1973a), employing this same technique, was able to assign the upper 400 m of the water column as the area of feeding by tuna. The amount of labour involved in such analyses is illustrated by a similar study (Roger, 1973b) involving the food of pelagic euphausids; this work required the analysis of the stomach contents of 18,620 specimens.

The gut-content analyses of smaller organisms such as copepods are not only more tedious but seem to contain greater bias. The food reported for nine genera of copepods by Mullin (1966) and for *Calanus* (Schrader, 1971) in a study of faecal pellet sedimentation suggests an inordinately important role for planktonic diatoms and silico-flagellates. Studies of the role of bacteria as a food resource have been made by differential plate counts of the bacteria found in the fore gut and hind gut of *Simulium* and *Chiconomus* (Baker and Bradnam, 1976), and in the gut of marine oligochaetes (Giere, 1975), and *Scylla serrata* (Hill, 1976). Gut-content analyses were also employed by Porter (1974), in an attempt to determine the relative importance of zooplankton to coral nutrition. He describes a method for the *in situ* extraction of coral gut contents.

Tracer methodology as applied to food-web analysis

Tracer methodology is potentially a rapid and sensitive technique with which to unravel marine food-webs experimentally and short-term experiments provide the means to obtain quantitative rate estimates. This technique demands only that a label can be fixed to the food resource of interest, and that the label can subsequently be observed in a consumer of the food resource. Labelled food or prey species can be introduced to the community of interest in such small amounts that the resultant feeding will not be affected by the experimental introduction of the labelled food resource. In a number of cases it is possible to conduct such experiments *in situ*, thus employing a nearly undisturbed community. This permits one to observe the fate of the labelled food resource in the presence of all the alternate, naturally occurring food resources.

The following discussion of the application of tracer methodology to marine food-web analysis has been subdivided into the classes of food resources to be employed. This was done because the techniques required to introduce label differ markedly between the several food resource classes.

Tracer introduced as dissolved organic matter

Qualitative evidence for the uptake of dissolved organic matter (DOM) by marine organisms has frequently been obtained by incubating specimens with a single labelled compound or a defined mixture of labelled compounds. The ability of the test organism to utilize the added compound(s) is judged by the rate at which the organism incorporates radioactivity during incubation with trace quantities of the radio-chemicals. The role of DOM as a food resource for a number of marine invertebrates has been evaluated by Stephens (1960, 1962, 1963, 1964), Stephens and Schinske (1961), and Stephens and Virkar (1966) using ^{14}C-labelled glucose or individual amino acids. Sorokin and Vyshkvartsev (1974) observed DOM utilization by a wide variety of marine invertebrates in experiments employing ^{14}C-labelled algal protein hydrolysate. In other studies (Williams et al., 1976) reconstituted algal or defined mixtures of labelled amino acids have been employed.

With regard to employing mixtures of labelled amino acids one should note that Schlichter (1978), in studies of amino acid uptake by *Anemonia sulcata*, reported reciprocal inhibition between members of amino acids derived from the aspartate and glutamate families. Other groups which have been tested for DOM incorporation in tracer studies have been reviewed by Sepers (1977). Incorporation of DOM has been demonstrated for at least one member of the taxa Foraminifera, Protozoa, Porifera, Coelenterata, Rhynchocoela, Sipunculoidea, Bryozoa, Chaetognatha, Annelida, Mollusca, Echinodermata, Hemichordata and Pogonophora (Stephens and Schinske, 1958, 1961: Stephens and Kerr, 1962; Southward and Southward, 1970; Testerman, 1972; Sorokin and Vyshkvartsev, 1974; Smith and Wiebe, 1977).

Technique designed for preparation of naturally occurring DOM

Ideally, such experimental food-web analyses should employ tracer in the chemical species of DOM currently being circulated in the community under investigation. To date, this has been possible only for those compounds which are formed by photosynthetic fixation and then subsequently excreted by autotrophs. The uptake of ^{14}C-labelled DOM by *Crassostrea gigas* larvae was studied by Fankboner and De Burgh (1978) by employing the labelled exudate from the brown macrophyte *Nereocystis luetkeana*. The labelled exudate was prepared by isolating terminal blades in clear polyethylene bags and incubating them *in situ* for 24 h in seawater containing 31 kBq . ml^{-1} of NaH^{14}CO$_3$. The labelled DOM preparation was freed of labelled particulate organic carbon (POC) by membrane filtrations (0·2 μm, pore size) and freed of the added radioactive dissolved inorganic carbon (DIC) by acidification (pH = 2·5) with concentrated HCl. After being left to stand for several hours to drive off ^{14}CO$_2$ the preparation was neutralized (pH = 7·9) with NaOH, kept frozen until the day of an experiment, and filtered through a membrane filter prior to use.

The incorporation rate of DOM by the microheterotrophic constituents of marine-water samples was estimated by Wiebe and Smith (1977a), employing ^{14}C-labelled DOM excreted by a naturally occurring phytoplankton population. Highly radioactive DOM preparations (2·5 kBq . ml^{-1}) were obtained in short-term incubations (50–80 min) by first concentrating the phytoplankton in seawater having all the naturally occurring DIC replaced by carbon-14 (22·7 mg^{14}C . l^{-1}).

Tracer introduced as labelled primary producers

Identification of the herbivorous members of a marine community can be obtained by those techniques employed to investigate bio-amplification of pesticide residue during trophic-level transfers. In such studies unialgal cultures have been labelled by exposure to low concentrations (10 μg . 1^{-1} of radioisotopically labelled halogenated organic insecticides (Nimmo *et al.*, 1970; Wheeler, 1970; Rice and Sikka, 1973*a*, 1973*b*; Petrocelli *et al.*, 1975*a*, 1975*b*). Subsequent measurements of radioactivity in the heterotrophic members of the community provide evidence of grazing upon the autotrophic members plus the transfers to higher trophic levels.

Algal cultures, labelled with photosynthetically fixed carbon-14, have frequently been employed to estimate feeding rates and afford a direct means of identifying the grazers in a community by merely assaying organisms for the presence or absence of label. Relative grazing rates of *Calanus helgolandicus* on large and small cells of *Ditylum brightwellii* were obtained in this manner by Richman and Rogers (1969). Experiments more akin to those envisaged for *in situ* food-web analyses are those reported by Porter (1976) who studied the grazing on labelled *Ankistrodesmus falcatus* in the presence of unlabelled cells of *Sphaerocystis schroeteri*. In similar experiments McMahon and Rigler (1963) employed a variety of foods labelled with ^{32}P. Labelled phytoplankton populations afford a more realistic resource in food-web analysis and have been employed in studies of secondary production (Chymr, 1967; Shushkina and Sorokin, 1969), the trophic position of single species (Chang and Parsons, 1975) and the *in situ* food-web structure of demersal zooplankton populations (Smith *et al.*, 1979).

Specialized preparation techniques

Highly radioactive preparations of ^{14}C-labelled phytoplankton were obtained during the preparation of labelled DOM by the method of Wiebe and Smith (1977). During a 50-minute incubation the phytoplankton obtained from 4 litres of seawater had incorporated in excess of 1·6 MBq. Photosynthetic fixation of radioactivity can be achieved by the light fixation of ^3H_2O as well as $DI^{14}C$ (McKinley and Wetzel, 1977) but due to the high concentration of ^1H_2O in seawater (*c.* 55 M) relative to the $DI^{12}C$ (2 nM) concentration, much larger quantities of label are required. Initial radioactivities of 160 Bq/cell have been obtained during 5-minute incubations of diatom auxaspores in ^3H_2O-labelled seawater (Smith *et al.*, 1979); however, 10 ml of seawater containing 37 GBq as ^3H_2O were required.

The distribution of naturally occurring elements (Young, 1970) and introduced stable isotopes, as well as radionuclides, have been employed as tracers in studies of marine phytoplankton. Intracellular turnover rates of carbon and nitrogen have been estimated using ^{13}C and ^{15}N (Slawyk *et al.*, 1977) and silicate uptake has been estimated in experiments by employing ^{30}Si (Nelson and Goering, 1977). The advantage of radioisotopes over stable nuclides remains, however, their relative ease of assay and the much greater sensitivity of measurement obtainable.

Of special interest is a method of radioisotope counting which is non-destructive to living samples. If sufficiently energetic isotopes are employed the radioisotope content of a living sample in seawater can be estimated by monitoring the Cerenkov radiation. Using this technique Pomeroy et al. (1974) were able to monitor the flux of ^{32}P in coral specimens over a period of several days without causing their death or apparent injury.

Tracer introduced as labelled secondary producers

The preparation of labelled secondary producers (we use the term to encompass all heterotrophs) has been accomplished by introducing label directly into the species of interest or by feeding them prey which has been directly tagged. As the techniques of preparing the labelled food, required by the latter approach, are covered in other parts of this section, we limit the discussion to methods of direct tagging.

The evaluation of conspecific predation on fry by adult *Cichlasoma nigrofasciatum* was studied by FitzGerald and Keenleyside (1978) employing ^{131}I as a label. Radioiodide (37 MBq) as glyceryl trioleate was ground with 1 ml of egg yolk and fed to the fry. After a feeding period of 24 h samples of 15 fry contained approximately 13 Bq which subsequently were lost at a turnover rate given by a biological half-life of about 24 hours. Larval *Micropterus salmoides* were labelled by a seven-day incubation in aquarium water to which ionic ^{85}Sr had been added (200–600 kBq . l^{-1}) (Carlson and Shealy, 1972). Net radioactivities as high as 10 kBq per specimen were obtained and the biological half-life of the ^{85}Sr in postlarval specimens was approximately 45 days. Similar incorporation patterns were obtained by Carter and Nicholas (1978) who studied the turnover of zinc in the aquatic larvae of *Simulium ornatipes*. Radioactivity from $^{65}Zn^{2+}$ (37 Bq . ml^{-1}) was incorporated into the larvae during a 24-hour incubation and subsequently washed out at a rate equivalent to a biological half-life of approximately 16 hours. The incorporation of tritium into the organic compounds of *Artemia salina* hatched and reared in seawater containing $^{3}HO^{1}H$ (19 MBq . m^{-1}) has been studied by Higuchi and Mukade (1976). After five days of growth the tritium concentration of the lyophilozed nauplii tissue had risen to within 40 per cent of the concentration of the exogeneously supplied ^{3}H-O-^{1}H.

Discrete size fractions of demersal zooplankton were labelled by brief incubations (5 minutes) in seawater containing (4 GBq . ml^{-1}) of $^{3}H_2O$ (Smith et al., 1979). The labelled prey were employed during *in situ* experiments designed to observe predation by size classes of zooplankton other than the prey size class. Although the biological half-life of the $^{3}HO^{1}H$ incorporated into the animals was short (8 h) the initial activities were high, e.g. for *Oithona* sp. approximately 2 kBq per individual.

100

Tracer introduced as labelled detritus

For the purpose of this discussion, we define detritus as consisting of both non-living particulate organic matter (POM) and the associated microbial flora. Indeed, given the low N/C ratio of the former, the latter is probably the most important constituent of detritus. The role of micro-organisms in the transfer of detritus through the food-web has been demonstrated in the altered rates of radioisotope incorporation in the presence and absence of antibiotics. Harris (1957) found that antibiotics inhibited the incorporation of ^{32}P-labelled phosphate by *Gammarus* as did Rigler (1961) working with *Daphnia magna* and Johannes (1964) in studies employing a marine amphipod. For reviews of the role of micro-organisms in transfer of radionuclides through food-webs see Peroni (1970) and Sieburth (1976).

The techniques of preparing labelled detritus are those previously discussed under tagging of primary and secondary producers. The labelled biota is harvested and converted to a physical and chemical form consistent with the system under investigation.

Fluorescent dye as a tracer

Not all labels used in food-web analysis are radioactive. Lane *et al.* (1976) stained prey organisms with acridine orange, and released them into a grazing chamber suspended in the water column. The guts of predators were later examined for evidence of fluorescence. The technique was first used in lakes, but is equally applicable in salt water. In some experiments Rhodamine B appears to be a superior stain, and freezing is a better method of preservation than that used originally (Lane, 1978).

3.1.3 MEASURING RATES OF CYCLING OF ELEMENTS[1]

Introduction

Only a few decades ago the concept of measuring the metabolic activity of a community would have been met with a mixture of horror and disbelief by most workers engaged in studies of metabolism. Even today the suspicion lingers, 'that really nice people' deal only with defined media, axenic cultures, or reproducible enzyme preparations. Ironically, at the end of the last century, the fledgling study of biochemistry was unsuccessfully fighting the inverse criticism of physiologists, that one cannot extrapolate from events observed in a test tube to those taking place in the living organism. In the end, the biochemists agreed in theory, declared themselves to be studying *in vitro* reactions, and it remained for the as-yet-unborn group, molecular biologists, to wed once again biochemistry and organismic physiology.

Students of community metabolism have, in the past, fought the same

1. By D. F. Smith.

101

battle as the earlier biochemists, but from opposing camps. The growing acceptance of the validity of community metabolic studies has, in part, been forced by the realization that the superposition principle is simply inapplicable to most ecosystems. This is hardly surprising since the application of superposition requires that one treat only small changes occurring in linear systems; ecosystems can be characterized as having variables which take large excursions and in possessing numerous non-linearities.

Given that superposition does not apply to ecosystems one must forego experiments using isolated species and instead conduct measurements on complex ensembles, preferably in the field. This restriction does not imply that such measurements will lack the rigour or even the reproducibility obtained in laboratory studies. It should not be forgotten that an English parson, observing the relationships between the reactants and products of microbial respiration in the marshes near his home, formulated the atomic theory.

This section deals with techniques currently employed to estimate the rates of chemical transformation of the biologically ubiquitous elements, carbon, oxygen, nitrogen, phosphorus, and sulphur. Special attention is given to those techniques which can be routinely applied in field studies and to those which least modify the system being studied.

Techniques for estimating the rates of carbon cycling:
carbon reduction rates

Estimates of the rates of particulate organic carbon (POC) production by photosynthesizing phytoplankton are most frequently made using a modification of the carbon-14 technique introduced by Steemann-Nielsen (1952). This technique involves the *in situ* incubation of replicate seawater samples in light and dark bottles after the addition of $NaH^{14}CO_3$. The samples are carefully filtered (Berman, 1973), the filters freed of $NaH^{14}CO_3$ (Sournia and Ricard, 1976), then dried and the radioactivity present in POC is counted. The radioactivity incorporated by the sample is converted to equivalents of carbon by correcting for sample quenching and independent estimation of the sample dissolved inorganic carbon (DIC) concentration (Smith and Wiebe, 1976).

Earlier estimates of photosynthetic rates were obtained by measuring the oxygen evolution in light versus dark bottles. This method has been compared with the ^{14}C-technique by McAllister *et al.* (1961) and Thomas (1964) who measured oxygen concentrations by the Winkler method and by Müller and Knopp (1971) and Littler (1973) who employed oxygen electrodes.

Primary production estimates obtained by the Steemann-Nielsen ^{14}C-method were consistently higher than those obtained by oxygen measurements. Oxygen measurements made at high oxygen concentrations by the Winkler method were in good agreement with those obtained using oxygen electrodes. At low oxygen concentration oxygen electrodes yielded estimates with higher precision than those obtained by the Winkler technique.

The employment of pH measurement to estimate carbon fixation rates in light versus dark bottles has been reported to be more reliable than oxygen determinations (Thomas, 1964; Littler, 1973).

An alternative approach to *in situ* incubation, yet one which is reported to estimate *in situ* phytoplankton production, has been advocated by Jitts *et al.* (1976). This technique involves laboratory incubation of water samples at various irradiance levels and the *in situ* monitoring of submarine irradiance. The construction of photosynthetic rate versus irradiance curves and the submarine irradiance measurements furnish sufficient information to estimate *in situ* production. The attractiveness of this approach is the relative ease with which *in situ* submarine irradiance measurements can be made in comparison with *in situ* sample incubations.

The discussion to this point has assumed that all the photosynthetically fixed carbon is to be found on the filter as POC. In fact, a varying percentage (1 per cent to 50 per cent) appears as dissolved organic carbon (DOC) in the filtrate. While the exact composition of DOC remains unsettled and the rate of production relative to that of POC varies (Chrost and Wazyk, 1978), release of DOC is currently considered to be important in quantitative studies of primary production and trophic relationships (Derenback and Williams, 1974; Chrost and Wazyk, 1978). Details of the techniques required to estimate the rate of DOC production are to be found in the literature (Smith and Wiebe, 1976; Wiebe and Smith, 1977a). Special care is required in technique if DOC production rate estimates are to be meaningful. Employment of too high a vacuum during filtration can result in cell lysis (Berman, 1973), and employing $NaH^{14}CO_3$ not obtained by collecting volatilized $^{14}CO_2$ or not sparging the acidified filtrate with carrier $^{12}CO_2$ can result in filtrate radioactivity which does not originate from DOC (Smith and Wiebe, 1976).

The incorporation of DOC into POC by the mediation of the microheterotroph population has been investigated by four methods: (a) Parsons and Strickland (1962), Wright and Hobbie (1965, 1966) and Crawford *et al.* (1974) added specific ^{14}C-labelled organic substrates to plankton samples and examined incorporation and dissimilation; (b) Azam and Holm-Hansen (1973) and Derenback and Williams (1974) modified this 'heterotrophic potential' technique by separating microheterotrophs from autotrophs by differential membrane filtration after $^{14}CO_2$ incubation in the light; (c) Derenback and Williams (1974) also measured DOC metabolism by monitoring the loss of DOC from the medium when plankton samples were incubated in the dark subsequent to their being incubated in the light with $^{14}CO_2$; (d) Wiebe and Smith (1977a) employed the phytoplankton in one aliquot of a seawater sample to produce labelled DOC which was subsequently added to a second aliquot that was used to estimate DOC incorporation by the microheterotroph population. For a recent review of incorporation of DOC by aquatic communities see Sepers (1977).

Although the techniques described above are general in application, a number of specific techniques have been developed to deal with benthic primary production. Littler and Murray (1974) estimated the primary productivity of marine macrophytes by measuring oxygen production. Light and dark control bottles containing representative branches or blades from different thalli were incubated at the ambient seawater temperature for up to 5–6 hours. After sampling the water for oxygen determinations the macrophytes were returned to the laboratory and an image of each was obtained by photocopying to normalize productivity estimates to a unit area of thallus.

Productivity estimates for *Laminaria* species have been made by Mann (1972) by estimating the growth rate of the blades. Small holes were punched 10 cm from the junction of stipe and blade and the distance the holes moved away from the stipe was recorded at intervals. By constructing length versus weight curves it was possible to convert length increment to biomass production.

The *in situ* turnover rates of POC were estimated by Smith and Wiebe (1977) for *Marginopora vertebralis*, a large benthic foraminiferan containing symbiotic zooxanthellae. Individuals were labelled by a short term $NaH^{14}CO_3$ incubation in the light in seawater containing no $NaH^{14}CO_3$. After the incubation the specimens were returned to the benthic substrate from which they had been taken and specimens periodically retrieved for assaying the radio-isotope content of the POC.

Marshall *et al.* (1971) employing $^{14}CO_2$ fixation studied the photosynthetic rates of marine benthic diatoms in undisturbed sediment samples. After incubation, sediment samples were filtered, burned in a sample oxidizer, and the $^{14}CO_2$ content assayed in a liquid scintillation counter. Eaton and Moss (1966) had obviated the problem of dealing with sediment by trapping the vertically migrating diatoms on lens tissue and incubating suspensions of the diatoms with $NaH^{14}CO_3$.

Estimation of carbon oxidation rates

Two major approaches to the problem of measuring community respiration are commonly employed. The choice of which to use is dictated in part by the question being asked, and in part by the character of the system being studied.

The first commonly employed strategy is to isolate the community of interest from the surrounding environment (i.e. prevent oxygen exchange). Changes in oxygen content, O_2, are monitored with time by analysing discrete water samples or by continuous monitoring; from these time-varying data the rate of net oxygen consumption is calculated.

This approach, with its principles, techniques, and restrictions, has been described by Odum and Hoskin (1958), Odum *et al.* (1959), Kinsey (1972) and Sournia (1976). The technique involves isolating the community of interest; aquatic communities can be enclosed in plastic bags, benthic communities can be covered by glass jars (Sournia, 1976).

The isolated communities are then kept in total darkness by shielding transparent containers with black plastic or aluminium foil or by employing opaque containers. This step in the experimental protocol introduces two complicating factors. First, the community metabolism, including respiration, may have different rates in the dark than in the light. This effect could be large and is notoriously difficult to evaluate. Second, shielding can cause large temperature changes, therefore sample temperature must be monitored to ensure this does not occur. Assuming dark and light respiration rates are equivalent, the community respiration rate is calculated from the time-varying oxygen concentration estimates which may be obtained by techniques described in the latter part of this section.

The second approach, which obviates the need to isolate the community of interest, has been described by the LIMER 1975 Expedition Team (1976), Smith and Jokiel (1975), and Pilson and Betzer (1973). This technique avoids the problems inherent in trying to isolate a community without altering the community metabolism, but requires careful monitoring of water movement, velocity, and depth. It is ideally employed in systems having a constant, uni-directional flow of water. The technique requires sampling along a transect defined by the path taken by a water mass. Simultaneous samples taken along the transect reflect changes due to the aquatic and benthic communities lying between the sampling stations. The rates of change are obtained from the ΔO_2 values and the interstation transit times of the water mass. A complete analysis of the community respiration pattern of such a system is obtained by sampling the transect over a suitable time period.

A third possible approach to estimate the rates of oxidation of organic carbon would be measurement of changes in dissolved organic carbon rather than in dissolved oxygen. Techniques by which to measure dissolved organic carbon concentration are given in the following section under the discussion of dissolved organic nitrogen measurement.

Techniques of oxygen concentration measurement

A variety of techniques are available for determining oxygen concentration and there are a large number of oxygen-determining instruments, employing physical and chemical methods, commercially available. The commonly used physical methods of oxygen determination employ the magnetic suscepti-bility of oxygen (paramagnetic techniques) or the redox potential of oxygen (polarographic techniques).

Paramagnetic oxygen determinations are based upon the fact that only oxygen of all the atmospheric gases is paramagnetic, the rest being diamagnetic. Thus measuring the magnetic susceptibility of a mixture of atmospheric gases is to estimate the concentration of oxygen in the mixture. The detailed pro-cedures for applying this principle to determine dissolved oxygen concentration, and a review of currently employed techniques, are to be found in a recently published symposium proceedings (Degn et al., 1976).

The polarographic method of oxygen determination depends upon the electrochemical reduction of oxygen occurring at a potential of 0·75 volts. The electrochemical redox reaction generates an electrical current in the mea-surement system which is proportional to the amount of oxygen in the sample. This technique has been employed by Kleinen Hammans et al. (1977) to study

105

the light-dependent oxygen uptake of the blue-green alga, *Anacystis nidulans*, by generating a polarographic action spectrum. The approach is relatively easy to employ and is frequently encountered in studies of respiration of zooplankton (Comita, 1968; Shushkina, 1972; Ikeda, 1974), teleosts (Saunders, 1963; Edwards *et al.*, 1972) and a variety of benthic forms (Mason, 1977).

Several other techniques deserve mention, as they have been employed in studies at low temperature or requiring especially high sensitivity. Respiration rates were measured at 20° C by Mason (1977) who employed an oxygen electrode system. This system proved to be unresponsive at 8° C and was discarded in favour of standard Warburg manometry.

Still another highly sensitive measure of respiration was required by Vernberg and Coull (1974) in their study of an interstitial ciliate. By employing a respirometric technique based on a Cartesian diver, as few as nine organisms were required for an experimental determination.

A technique which has been successfully employed in fields other than marine science employs the bacterial luciferase system (Hastings, 1955; Hastings and Gibson, 1963). The procedure is a modification of the one employed to estimate ATP with oxygen replacing ATP as the limiting reactant.

Chemical methods for the estimation of oxygen concentration include the reaction of oxygen with copper (Brooks *et al.*, 1952), the quenching of fluorescence of pyrenebutyric acid (Longmuir and Knopp, 1973), the binding of oxygen by haemoglobin (Hultborn, 1972) and the reaction of oxygen with alkaline pryogallol (Williams *et al.*, 1952).

A frequently employed technique, described by Strickland and Parsons (1965), is based on the oxidation of manganous to manganic hydroxide by oxygen followed by iodometric titration. This technique, first published by Winkler (1888), is still widely employed in marine studies (Nival *et al.*, 1972; Ikeda, 1974; Sameoto, 1976).

Techniques for estimating the rate of nitrogen cycling

Unlike the other elements discussed in this section nitrogen possesses no convenient radioactive isotopes which can be employed as tracers to estimate the kinetics of turnover. The employment of the heavy isotope ^{15}N does furnish one with a true tracer, but the sensitivity associated with radioisotopic labelling is lost. The employment of ^{15}N in community metabolism studies will be referred to at the end of this segment, but in the main, chemical assays furnish the methods of choice in estimating rates of nitrogen fixation and utilization of fixed nitrogen compounds.

Estimates of N_2 fixation rates

Rates of nitrogen fixation *in situ* are most commonly made using the acetyline-reduction technique, described by Stewart *et al.* (1967), which depends upon the fact that nitrogenase is capable of reducing acetylene to ethylene. Most assays involve partially filling a container with a water sample, capping the container and replacing the vapour phase with a synthetic gas mixture containing acetylene, carbon dioxide, argon and oxygen. Following a suitable incubation period a unit volume of the gas phase is isolated in a hypodermic syringe and the ethylene content assayed by gas chromatography (Webb et al., 1975; Bohlool and Wiebe, 1978).

The assay is sensitive and reproducible, but does demand careful experimental technique. Commonly occurring errors can be avoided by careful experimental design (Flett et al., 1976) and by taking a number of time-varying aliquots for ethylene assay (Wiebe et al., 1975).

Estimates of utilization rates of fixed nitrogen

Two approaches are commonly employed to estimate rates of utilization of fixed nitrogen compounds (FNC); both depend upon analyses of the concentration of the chemical species of interest. In the first of these, samples are incubated in seawater containing the FNC at a concentration very near or equal to that of the ambient seawater. Samples of the water are removed at noted intervals and the FNC concentration measured. A set of such data points, plotted against time, permits one to calculate the net rate of uptake or excretion of the FNC from the initial slope of the curve (Franzisket, 1975; Harvey and Caperon, 1976). The second approach requires a set of experiments, like the one just described, each conducted at a different initial FNC concentration (Cleland, 1967; Caperon and Meyer, 1972; Falkowski, 1975). The initial concentrations and the associated rates permit one to calculate values for K_s and V_{max}, parameters which may be formally analogous to those of the Michaelis-Menten equation.

Both of these approaches require that the observed rate estimates be normalized to a unit of the biological component producing the change. The choice of the normalization parameter is dependent upon the systems being studied. Crossland and Barnes (1974) employed both chlorophyll a and protein in work with Acropora, and Webb et al. (1975) measured surface area in studies employing algal pavement.

Both approaches also demand the measurement of the ambient concentrations of NO_3^-, NO_2^-, NH_4^+, and dissolved organic nitrogenous compounds (DON), regardless of which chemical species is the FNC of interest. The rate of uptake of an FNC is strongly influenced by the ambient concentrations of the other FNCs (McCarthy et al., 1977). A classic example drawn from microbiology is the preferential utilization of NH_4^+ if NH_4NO_3 is supplied as the sole nitrogen source. Only when the NH_4^+ is depleted does NO_3^- incorporation begin. A variety of chemical assays for each of the FNC's are currently being employed. Although the method of choice will be dictated, in part, by the system being studied, the following techniques are recommended.

107

Dissolved organic nitrogen is measured directly (Moore and Stein, 1954) as NH_4^+ following Kjeldahl digestion (Strickland and Parsons, 1972) or as NO_3^- following photo-oxidation (Armstrong et al., 1966). Primary amine-N can be estimated using fluorescamine (North, 1975), ninhydrin-positive substances and NH_4^+ by the technique of Snow and LeB. Williams (1971), and urea by a diacetyl-monoxime technique (Newell et al., 1967). Ammonium ion concentrations are commonly measured by the method of Solórzano (1969) employing phenolhypochlorite or a modification of this technique (Grasshoff and Johannsen, 1972). Nitrate-nitrite determinations frequently employ the Cd-Cu reduction column techniques of Wood et al. (1967).

While the employment of ^{15}N as a tracer has never become commonplace, it is more and more frequently encountered, and given the recent advances in laser isotope spectrometry, ^{15}N studies may soon proliferate. The basic techniques of ^{15}N measurement have been described by Barsdate and Dugdale (1965) and Desaty et al. (1969). The techniques of employment of ^{15}N in marine studies can be found in publications by Dugdale (1967), Dugdale and Goering (1967), Slawyk et al. (1977), McCarthy et al. (1977) and Wada et al. (1977).

Techniques for estimating the rates of phosphorus cycling

Before phosphorus isotopes became commercially available estimates of phosphorus cycling rates were obtained by measuring the rate of change in phosphate concentration. The incorporation of phosphate was estimated by measuring the rate of decreasing phosphate concentrations in the water by the method of Murphy and Riley (1962). This spectrophotometric assay has received subsequent modifications to increase its utility when employed to analyse seawater samples (Olsen, 1967; Strickland and Parsons, 1972).

Rates of formation of organic phosphorus from phosphate were accomplished by estimating the phosphate concentration of samples before and after the complete oxidation of all the phosphorus to phosphate. Oxidation was achieved either by the persulfate oxidation technique (Menzel and Corwin, 1965) or, more recently, by UV-photo-oxidation (Armstrong et al., 1966).

Tracer methodologies unique to phosphorus studies

Estimations of phosphorus turnover rates by chemical assays are subject to the same limitations discussed under previous headings in this discussion. Moreover, in contrast to nitrogen, phosphorus affords us two commercially available radioisotopes, [32]P (half-life: 14·3 days) and [33]P (half-life: 25 days). The availability of two isotopes with half-lives differing nearly twofold, and the fact that ^{32}P has a β-mode of decay with a decay energy of 1·71 MeV, affords experimental approaches to rate studies that are unique to studies of phosphorus.

Doubling labelling of a single element

The two isotopes of phosphorus, ^{32}P and ^{33}P, permit one to measure the flux from phosphate to organic phosphorus simultaneously and from organic phosphorus to phosphate in a single experiment. Furthermore, the differences in both the length of half-lives and in decay energies permit this simultaneous measurement to be accomplished in two quite different ways.

As an example, consider an experiment in which the dissolved phosphate in seawater is labelled with $H^{32}PO_4{}^{2-}$ while some of the organic phosphorus (dissolved organics, or bacteria, zooplankton, etc.) has been previously labelled with ^{33}P. One could estimate the rates of conversion of organic phosphorus to phosphate by removing samples of water and estimating the $H^{32}PO_4{}^{2-}$ content in the presence of $H^{33}PO_4{}^{2-}$.

This may be done, even if only a single channel counter is available, by counting the sample repeatedly at recorded times (time elapsed since the beginning of the experiment). Each time the sample is counted the observed dpm will increase because of the loss of radioisotope by decay. The observed dpm is the sum of the dpm arising from ^{32}P and ^{33}P and the observed dpm at various counting times is expressed as:

$$\text{dpm} = A_{32}e^{-14\cdot3t} + A_{33}e^{-25t}. \tag{105}$$

Thus by counting the sample on several different days the recorded values of dpm and t can be used to solve Equation (105) to obtain A_{32} and A_{33}, the quantities of ^{32}P and ^{33}P present at the time of the experiment. For details of the technique see Rescigno and Segre (1966).

The difference in the ^{32}P and ^{33}P decay energies permits one to measure both isotopes in the same sample simultaneously by standard dual-isotope-analysis techniques. This approach requires a liquid scintillation counter with three channels and, although far more rapid than the first technique, it is inherently less precise.

Finally, ^{32}P, because of its high decay energy, affords a unique opportunity to the investigator, the ability to monitor the radioisotope content of living organisms without destroying them. The high energy β-emission of a ^{32}P source in seawater causes light to be produced in the water by Cerenkov radiation, the same phenomenon seen as a blue glow around reactors immersed in water shielding. A labelled organism can be left in situ and periodically transferred to a counting chamber for radioactivity measurement. Details of this procedure are given by Pomeroy et al. (1974), who studied phosphorus exchange between reef water and living corals.

Techniques for estimating the rates of sulphur cycling

Sulphur is the final element that we will consider to be a major component of biological systems. Although present in smaller quantities than any of the elements previously discussed, it is found intracellularly as a constituent of macromolecules and low-molecular-weight metabolites and, in the environment, is a constituent of one of the most common routes of anaerobic metabolism.

The existence of a conveniently employed radioisotope of sulphur, ^{35}S, makes a tracer kinetic approach the method of choice in studies of sulphur cycling. The attendant chemical analyses which are commonly required and special approaches used in the analysis of sediment samples are treated below.

Quantitative analysis of sulphur compounds

Reduced sulphur compounds, like organic phosphorus and nitrogen compounds, may be estimated by assays for a specific chemical species or for bonds. More frequently, the total quantity of the element is assayed after oxidizing all the organic species to the same redox state, i.e. after converting all the sulphur to SO_4^{2-}. The organic sulphur in particulate samples can be quantitatively converted to sulphate by digestion with a mixture of concentrated HCl and HNO_3. Contaminating cations are removed from the solution by passage through a cation exchange resin and the sulphate is quantitatively precipitated by the addition of excess $BaCl_2$ (Vogel, 1961). Dissolved organic sulphur compounds are converted to sulphate, for analysis, by ultra-violet irradiation after the method reported by Armstrong et al. (1966).

The chemical estimation of sulphide in seawater or sediment samples is accomplished by acidification of the sample to liberate H_2S. Free H_2S is collected in a series of three wash tubes containing 5,5′-dithiobis (2-nitrobenzoic acid) (DTNB). The quantity of sulphide present is calculated from the absorbancy of the thionitrobenzoate at 412 nm (Kredich and Tomkins, 1966; Skyring and Chambers, 1979). The chemical estimation of sulphate is still done routinely by gravimetric analysis of the barium salt. Details of the procedure are to be found in APHA (1971).

A few points concerning the employment of ^{35}S in turnover studies should be noted, in particular a new rationale which has been described for sediment samples. In studies of sulphate reduction the isotope is introduced as $^{35}SO_4^{2-}$ and the rate of reduction estimated from the rate of appearance of label in H_2S. Either $H_2^{35}S$ is liberated from the sample, collected and counted (Chambers and Trudinger, 1975), or one measures the radioactivities of aliquots of thionitrobenzoate solutions produced during chemical estimations of sulphide (Skyring and Chambers, 1979).

Studies of sulphate metabolism are intrinsically associated with anaerobic environments, particularly sediments. In such a system, to slurry or agitate the core sample is to destroy its relationship to the sediment from which it originated. Skyring and Chambers (1978) have devised an assay in which $Na_2^{35}SO_4$ is dissolved in a solution of 2 per cent sodium silicate, which is used to coat a glass rod. The glass rod is dried and is employed to introduce $^{35}SO_4^{2-}$ to the sample by plunging it into a sediment core. The authors show that unequal distribution of $^{35}SO_4^{2-}$ throughout the core cannot influence the estimates of

110

rate of sulphate reduction if the entire core sample is used in the subsequent collection of $H_2{}^{35}S$.

3.1.4 COMPUTATION OF FLUXES[1]

The techniques given in the previous section provide a reasonably adequate basis for determining fluxes of the key elements when they occur in solution, or where their rates of movement into and out of organisms can be determined by tracer techniques. However, trophic transfers between organisms which are too large, or range over too wide an area for tracers to be used conveniently, present special problems. For example, measurement of the productivity of invertebrates in the benthos, and removal of that production by shoals of fish, does not lend itself to solution by the tracer methods given in Section 3.1.2. Nor has it proved possible to follow quantitatively the transfer of primary production by marine macrophytes into the detritus food-web on an estuary, using tracer techniques.

The key ecological flux between primary producers and herbivores depends on the rates of two processes: (a) generation of new biomass by the plants (net primary production); (b) removal of part of that biomass by grazers. In practice, most tracer or chemical measurements of community photosynthesis and respiration leave unanswered the question of how much net plant biomass has been produced. Wherever possible, direct measures of net biomass production should be used, as this gives an unambiguous measure of the amount of plant tissue made available to grazers (see Mann, 1969). Similarly, it is usually necessary to make independent estimates of secondary production by benthos and by zooplankton. Techniques for doing this are reviewed in Mann (1969) and in Crisp (1971). Direct measurement of secondary production for each species is difficult and time-consuming but there is an increasing body of knowledge of how production/biomass ratios vary according to the size and length of life of different organisms. Thus, the production of a benthic community can be estimated from data on annual mean biomass, together with knowledge of the sizes and life histories of the dominant organisms. Production is often expressed as g dry wt $m^{-2}y^{-1}$, and this can be converted to its equivalent content of the appropriate element (C, N, P, etc.) after analysis of samples of the material produced.

Removal of production by large predators or grazers can be approached in a number of ways. In the equation $P = E + \Delta B$, where P = net biomass production, E = removal by predators or grazers, and ΔB = biomass change, removal can be calculated by the difference between total net production and biomass accumulation. The other main method is to make a quantitative study of the feeding rates of the consumer organisms. For zooplankton and other small animals, feeding rates may be measured under controlled conditions in

1. By K. H. Mann.

field or laboratory, but for large fishes it is often necessary to resort to indirect methods, such as making a materials budget from separate measurements of growth, respiration, egestion and excretion, or by determining the number of times the stomach is filled in the course of 24 hours. These methods are reviewed in Gerking (1978).

Production by microbes

Pomeroy (1974) by separating drew attention to the fact that when micro-organisms below the size of net phytoplankton are separated from the plankton, their respiration often accounts for 50–90 per cent of the respiration of the total. Hence it seems possible that they are responsible for some very major fluxes in planktonic systems. Some of the problems of measuring fluxes through the microbiota have been touched upon in the discussion of tracers in Section 3.1.2 but the question of how much production of biomass is attributable to micro-organisms is an important yet difficult question. Sorokin (1969) gives a method which involves measuring uptake of labelled bicarbonate in the dark, and use of the assumption that this constitutes 6 per cent of newly synthesized bacterial carbon. Results obtained by this method have been challenged by Banse (1974). A recent discussion of the problem is found in Sieburth (1977). It seems probable that, in addition to bacteria, ciliates, rotifers and nauplii are important pathways of secondary production in some planktonic situations (Eriksson *et al.*, 1977).

3.2 Thinking in terms of scale: introduction to dimensional analysis[1]

Natural laws, when properly written in mathematical form, are equally valid whatever system of scientific units is used to express them. At a deeper level of generality than the mere units lie the fundamental *dimensions* of mass, length and time. To say that universal laws should be independent of the system of units is another way of saying that they should be *dimensionally consistent*: one should not equate ships with sealing wax, nor can one add cabbages to kings.

We can exploit the basic principle of dimensional consistency to find useful results from differential equations that are otherwise insoluble; to elucidate the laws describing the relative dynamics of natural systems and their small-scale replicas; and, if our intuition is good, to make quantitative predictions about processes so complex that we are unable to write down their governing equations. A useful elementary reference to the approach is Pankhurst (1964).

---

1. By T. Platt.

Non-dimensionalization of differential equations

As a simple example of the power of dimensional analysis, let us consider the classical red-tide problem first posed by Kierstead and Slobodkin (1953). A patch of water of size ξ contains phytoplankton growing at the exponential rate r. The environmental conditions are such that growth is impossible outside of the confines of the patch. Given that there exists a field of turbulence whose intensity is characterized by a diffusion coefficient K_H, we ask: what is the condition that the patch of phytoplankton will persist in time? The appropriate differential equation is

$$K_H \frac{\partial^2 B}{\partial x^2} + rB = \frac{\partial B}{\partial t}, \tag{106}$$

where B is the biomass of phytoplankton, t is time and x is the position co-ordinate. (Note: for simplicity this is a one-dimensional representation. Here the word *dimension* refers to one of the three space co-ordinates x, y and z, rather than the fundamental dimensions of mass, length and time. Yet another use of *dimension* belongs to everyday speech: we say, the dimensions of this page are 15.4 cm $\times$ 24 cm. In dimensional analysis we would prefer to use the word *scale* here, and say the page has a characteristic scale of 15.4 cm in the horizontal and 24 cm in the vertical.)

Kierstead and Slobodkin proceeded to solve Equation (106) explicitly to derive the condition for the persistence of the patch. We can arrive at essentially the same conclusion, without solving the equation, by applying the method of non-dimensionalization. To do this we transform the variables of Equation (106) as follows:

$$B = \beta B_*,$$
$$x = \xi x_*, \tag{107}$$
$$t = r^{-1} t_*.$$

The old variables B, x and t have been transformed to new variables B_*, x_* and t_*, where the stars indicate that the variables are dimensionless. Since the Equations (107) have to be dimensionally consistent, the dimensions of the left-hand side must now reside in the quantities β, ξ and r^{-1} on the right-hand side. These quantities are called the *characteristic scales* for B, x and t.

The choice of characteristic scales for a particular problem requires some insight. Here we have chosen for the horizontal scale the size of the patch, ξ, and for the time scale the inverse of the exponential growth rate, r^{-1}. Choice of a scale for the biomass, β, is less crucial, since it will be eliminated in what follows.

If we now rewrite Equation (106) in terms of the starred variables, we have

$$r\beta \frac{\partial B_*}{\partial t_*} = \frac{K_H \beta}{\xi^2} \frac{\partial^2 B_*}{\partial x_*^2} + r\beta B_*. \tag{108}$$

113

[Remark:
$$\frac{\partial B}{\partial t} = \frac{\partial(\beta B_*)}{\partial(r^{-1}t_*)} = r\beta \frac{\partial B_*}{\partial t_*}$$

and
$$\frac{\partial^2 B}{\partial x^2} = \frac{\partial}{\partial x}\left(\frac{\partial B}{\partial x}\right) = \frac{\partial}{\partial x}\left(\frac{\beta}{\xi}\frac{\partial B_*}{\partial x_*}\right)$$
$$= \frac{\partial}{\xi \partial x_*}\left(\frac{\beta}{\xi}\frac{\partial B_*}{\partial x_*}\right) = \frac{\beta}{\xi^2}\frac{\partial^2 B_*}{\partial x_*^2}\Bigg].$$

Dividing through Equation (108) by $r\beta$ gives

$$\frac{\partial B_*}{\partial t_*} = \frac{K_H}{\xi^2 r}\frac{\partial^2 B_*}{\partial x_*^2} + B_*. \tag{109}$$

The two terms on the right-hand side of Equation (109) have opposing effects on the persistence of the phytoplankton patch. The diffusion term tends to erode the patch and therefore limits its persistence; the growth term tends to prolong its existence. The values of the characteristic scales for which both terms have equal magnitude should supply the minimum condition that the patch persist. Equating the coefficients of these two terms gives us an estimate of the critical patch size ξ_c. That is

$$\xi_c \simeq \left(\frac{K_H}{r}\right)^{1/2}. \tag{110}$$

To within a factor of π, this is identical to the result obtained through exact solution by Kierstead and Slobodkin (1953). Although the procedure described above seems complicated on first acquaintance, the result can usually be written down at sight by someone familiar with the method.

In equations with many terms we can apply a similar approach to establish the conditions under which particular terms might dominate the equation. An example is the general equation describing the rate of change of phytoplankton abundance given by Platt and Denman (1975). The equation includes terms for horizontal advection, vertical transport, sinking of cells, horizontal and vertical diffusion, light-dependence of primary production and phytoplankton-dependent grazing rate of zooplankton.

By non-dimensionalizing the equation through the choice of suitable scale factors, and comparing coefficients of terms, it is possible to see the circumstances under which, for example, the vertical velocity term might dominate the equation (an upwelling system, say), or the light limitation might be more important. O'Brien and Wroblewski (1973) have analysed the conditions under which advection is important in phytoplankton models.

Dimensional analysis applied to process equations

We can often apply dimensional reasoning to extend our scientific insight into complex dynamic processes. As an introduction to the approach consider first a purely physical problem (a related biological problem is treated below). Kolmogorov (1941) sought a theory of turbulent eddies of a size intermediate between the large-scale (where energy is put into the sea) and small-scale (where it is dissipated by viscous friction). This intermediate scale is called the inertial range, and it was postulated by Kolmogorov that in the inertial range, eddies transmit their energy from larger to smaller scales in such a way that the power spectral density, $S(k)$ (velocity variance per unit wave number), depends only on the wave number itself, k, and the rate, ε, at which energy is dissipated at fine scales by viscosity.

Let us denote the dimensions of a quantity by square brackets [ ]. Then the dimensions of $S(k)$ are

$$[S(k)] = \left[\frac{L^2}{T^2 \cdot L^{-1}}\right] = \left[\frac{L^3}{T^2}\right],$$

where L refers to length and T refers to time. The quantity $[L^2/T^2]$ refers to the velocity variance and $[L^{-1}]$ to the wave number. The energy dissipation rate (per unit mass of water) will have dimensions of velocity squared divided by time: $[\varepsilon] = [L^2 T^{-3}]$. With this as background we can now examine the consequences of Kolmogorov's hypothesis that the only relevant quantities are ε and the wave number. If this hypothesis is valid, what does it imply about the form of the spectral density $S(k)$?

Suppose that we can write $S(k)$ as a product of power functions of ε and k (it turns out that we are always allowed to do this):

$$S(k) \sim \varepsilon^p k^q. \tag{111}$$

Writing in the dimensions of both sides,

$$[S(k)] = [L^3 T^{-2}] = [\varepsilon^p k^q] = [(L^2 T^{-3})^p (L^{-1})^q] \tag{112}$$

or
$$L^3 T^{-2} = L^{2p-q} T^{-3p}. \tag{113}$$

Comparing indices on either side of Equation (113) gives us two simple, simultaneous equations for p and ε,

$$3 = 2p - q$$
$$-2 = -3p \tag{114}$$

with solutions $p = 2/3, q = -5/3$.

Thus, the following functional form for $S(k)$ is consistent with Kolmogorov's hypothesis:

$$S(k) \sim \varepsilon^{2/3} k^{-5/3}. \tag{115}$$

The prediction is that the distribution of velocity variance over length-scale (wave number) in this turbulent field should vary as the minus-five-thirds power of the wave number, a conclusion that was verified in the field by Grant *et al.* (1962). In other words, a testable prediction was developed through dimensional arguments for a complex problem for which a complete and exact solution could not be found. To summarize the method in this case, it consists of identifying the relevant variables; expressing the unknown function as a product of powers of these variables; writing the dimensions of each side of the equation and comparing indices.

Another application of dimensional analysis is to find suitable scale transformations such that physiological data from a variety of sources may be compared. An example is in the work of Palaheimo and Dickie (1965) who sought to scale the growth equation for fishes and compare data on fish of different ages, sizes, species, and feeding regimes. The data on metabolic rate were first adjusted for temperature differences using the standard Krogh correction. The basic empirical growth equation is

$$\frac{\Delta w}{\Delta t} = R e^{-(a+bR)}, \tag{116}$$

where w is the weight, R is the ration and a, b are constants. Differentiating the left-hand side with respect to ration, we see that growth will be maximal at $bR = 1$; it is therefore convenient to define the dimensionless ration R_* as

$$R_* \equiv bR. \tag{117}$$

What is the corresponding body weight? To see this, we write out the fundamental energy equation (the Winberg equation) of metabolism = ration − growth, i.e.

$$\alpha w^\gamma = R - \frac{\Delta w}{\Delta t} \tag{118}$$

with α and γ constants. Substituting from Equation (116)

$$\alpha w^\gamma = R(1 - e^{-(a+bR)}) \tag{119}$$

where $R_* = 1$, $R = 1/b$ and we find the required weight-scale to be

$$\left(\frac{1 - e^{-(1+a)}}{\alpha b} \right)^{1/2}.$$

This scale contains the factor e^{-a}, which corresponds to the maximum efficiency $(1/R)(\Delta w/dt)$ at low feeding levels. Paloheimo and Dickie chose to keep this factor separate from the weight-scale so that the effect of different feeding efficiencies could be presented more clearly. The dimensionless weight-scale was then defined as

$$w_* = (\alpha b)^{1/\gamma} w. \tag{120}$$

It remains to find the appropriate time-scale. We put the scale variables into Equation (118):

$$\frac{(\alpha b)^{1/\gamma} \Delta w}{b^{-1}(\alpha b)^{1/\gamma} \Delta t} = \frac{\Delta w_*}{b^{-1}(\alpha b)^{1/\gamma} \Delta t} = R_* - w^\gamma{}_*,$$

whence we deduce the dimensionless time-scale t_* to be

$$t_* \equiv b^{-1}(\alpha b)^{1-\gamma}t. \tag{121}$$

Then choices of scales for ration, weight and time were used successfully to affect a standardization of available data on fish growth such that general principles could be sought through a comparison of the results of diverse feeding studies.

Let us now return to the problem of estimating the functional form of a power spectral density, this time for the variance of chlorophyll concentration in the sea. The problem has been treated using dimensional arguments by Denman and Platt (1976) and using modern fluid dynamics by Denman *et al.* (1977).

Consider the chlorophyll to be a passive (but non-conservative) contaminant of a field of turbulent eddies with characteristic length-scale d and time-scale τ, the time within which the eddy transfers its kinetic energy to eddies of diameter $d/2$. The argument consists in comparing this time-scale τ with the time-scale r^{-1} for phytoplankton growth. When $\tau \ll r^{-1}$, the eddies will be eroded so fast that the spectrum of chlorophyll variance should resemble that of a conservative passive scalar; when $\tau \gg r^{-1}$, the time-scale is long enough for effects of biological growth to be apparent, and the spectrum might be expected to diverge from that of a conservative passive scalar.

Through arguments similar to those developed above for the velocity spectrum, dimensional analysis gives for the spectrum of a non-growing passive scalar such as temperature, θ,

$$E_\theta(k) \sim \chi_\theta \varepsilon^{-1/3} k^{-5/3}, \tag{122}$$

where χ_θ is a constant expressing the rate of destruction of temperature variance at small length-scales.

For the chlorophyll spectrum we include among the pertinent quantities a constant χ_β, analogous to χ_θ, but referring to β, the chlorophyll concentration. We also include the growth rate r. Applying the principle of dimensional homogeneity we find

$$[E_\beta(k)] = [r^a \chi_\beta^b k^c \varepsilon^d], \tag{123}$$

where a, b, c and d are constants to be determined. Then

$$[\beta^2 L] = [(T^{-a})(\beta^{2b} T^{-b})(L^{-c})(L^{2d} T^{-3d})]. \tag{124}$$

Note that we are now including the biomass β as a separate and independent dimension to the analysis in addition to the usual mass, length and time. Note also that since the energy dissipation rate ε is referred to unit mass, the mass appears on neither side of this equation. We therefore have only three fundamental dimensions (biomass, length and time) in Equation (124) and four unknowns (a, b, c and d). The problem is therefore not completely determinate.

Equating indices on both sides of (124) gives

$$2 = 2b$$

$$0 = -a - b - 3d \qquad (125)$$

$$1 = 2d - c$$

with solutions $b = 1$; $c = 2d - 1$; $a = -1 - 3d$. The indeterminacy is the reason that the solutions contain the unknown d. Equations (125) imply that we can write $E_\beta(k)$

$$E_\beta(k) \sim \chi_\beta r^{-1} k^{-1} (\varepsilon^d k^{2d} r^{-3d}). \qquad (126)$$

Taken together, the three terms in d form a dimensionless group and in general there will be as many such dimensionless groups as there are undetermined indices (Buckingham's theorem).

Another way to write (126) is

$$E_\beta(k) \sim \chi_\beta r^{-1} k^{-1} F\left(\varepsilon \frac{k^2}{r^3}\right), \qquad (127)$$

where F is an unspecified function of the dimensionless group $\varepsilon k^2/r^3$. To go beyond this result, we rely on our insight that, as in the Kierstead and Slobodkin problem, an important length-scale will be that for which growth and turbulent dissipation are just balanced. Looking at the structure of the dimensionless group $\varepsilon k^2/r^3$, and expressing this characteristic length-scale as a wave number k_c, we can construct it as

$$k_c \equiv (r^3/\varepsilon)^{1/2}. \qquad (128)$$

At wave numbers large compared to k_c, we expect that the fields are changing too fast for the biology to have an effect, and $E_\beta(k)$ should follow the temperature spectrum $E_\theta(k)$. At wave numbers small compared to k_c, we expect the β-spectrum to diverge from the θ-spectrum because the effect of growth cannot be ignored. If we make the further assumption that in this range of k, the time-scale for growth, r^{-1}, is the *only* relevant quantity, i.e. that ε is negligibly important, $E_\beta(k)$ is no longer indeterminate and dimensional analysis gives

$$E_\beta(k) \sim \chi_\beta r^{-1} k^{-1}. \qquad (129)$$

Again, we have a result that is testable in the field; the chlorophyll spectrum should go as k^{-1}. Relevant field data are given in Denman and Platt (1976), in Horwood (1978) and in Lekan and Wilson (1978).

A not dissimilar result was reached by Denman et al. (1977) through solution of the basic partial differential equation. But note that even in this treatment (Equation (10) of their paper) it was necessary to invoke dimensional arguments to estimate the time taken for chlorophyll variance to be transferred from one length-scale to another in the turbulent cascade.

Scale models and microcosms

To examine the potential application of dimensional analysis to ecological microcosms, let us first consider a simple example already given, that of the scaled form of the Kierstead-Slobodkin equation (Equation (109)):

$$\frac{\partial B_*}{\partial t_*} = \frac{K_H}{\xi^2 r} \frac{\partial^2 B_*}{\partial x_*^2} + B_* .$$

Now let us suppose that we have two systems: the natural system labelled by subscript 1 and the microcosm labelled by subscript 2.

We define a series of *scale factors* f_L, f_T and f_B to be the relative magnitudes of these variables in the two systems:

$$x_2 = f_L x_1,$$
$$t_2 = f_T t_1, \tag{130}$$
$$B_2 = f_B B_1 .$$

These scale factors are just the ratios of the characteristic scales used to non-dimensionalize the governing equations of the two systems. For example, for horizontal distance, $f_L = \xi_2/\xi_1$, so that

$$\frac{x_{1_*}}{x_{2_*}} = \frac{\xi_2}{\xi_1} \frac{x_1}{x_2} = 1$$

or
$$x_{1_*} = x_{2_*}, \tag{131}$$

and similarly for t_* and B_*. In other words, the magnitudes of the non-dimensional variables are the same in both systems. The implication, from Equation (109), is that the dynamics of the two systems will be exactly similar if the value of the dimensionless group $(K_H/\{\xi^2 r\})$ is identical in them both. (Note that boundary conditions and initial conditions have to be specified in scaled form.) To put it another way, the dimensionless biomass B_* will be an identical function of dimensionless space and time (x_*, t_*) in both systems, provided that these systems have equivalent values of the dimensionless quantity $(K_H/\{\xi^2 r\})$. In working with a dynamic simulation model of such a system, it is necessary to vary only the magnitude of this dimensionless group to see the whole range of possible behaviour of the system.

Now the idea of a microcosm certainly implies that the geometric scale ξ will differ in the two systems. For the dynamics to be similar, we would then have to make a corresponding change in (K_H/r). Because we are dealing with living cells, our latitude in adjusting r is strictly limited. Indeed, we might insist that the scale-factor for time between the two systems be equal to unity, i.e. that the cells divide at the same rate in the microcosm as they do in the real world. In this case, the effect of the geometrical scaling would have to be compensated for by changes in the diffusion coefficient alone. (It is of only limited help that K_H is scale-dependent (Okubo, 1971), since the scale dependence is not strong enough, by itself, to compensate for the factor of ξ^2.)

119

This highly simplified example begins to bring into focus some of the difficulties inherent in the application of the principle of physical similitude to systems involving biological entities. The problems are even more apparent when more realistic (and therefore more complex) systems are examined. For example, consider the introduction of just two additional factors; the light-dependence of phytoplankton growth rate and the acknowledgement of the vertical dimension. Following Platt et $al.$ (1977) we can represent this system by

$$\frac{\partial B}{\partial t} + \vec{u} \cdot \vec{\nabla}_H B + w\frac{\partial B}{\partial z} = K_H \nabla_H^2 B + K_v \frac{\partial^2 B}{\partial z^2} + P_m^B \left(\frac{\alpha I_z}{P_m^B + \alpha I_z}\right) \quad (132)$$

where $\vec{u}$ is horizontal velocity, w is vertical velocity, K_H and K_v are horizontal and vertical diffusion coefficients, I_z is the available light at depth z, P_m^B is the maximum specific growth rate at saturating light intensity and α is the initial slope of the light saturation curve.

Equation (132) may be scaled with the following set of transformation equations:

(Remark: Note that the scaling of horizontal velocity is not independent of the scaling of horizontal and vertical length, reflecting the fact that the dimensions of velocity are distance divided by time.)

$$
\begin{aligned}
B &= \beta B_* \\
x &= \xi x_* \\
z &= \Psi z_* \\
t &= (P_m^B)^{-1} t_* \\
w &= (v - \upsilon)w_* \\
\vec{u} &= v\xi\chi^{-1}\vec{u}_* \\
I_z &= P_m^B \alpha^{-1} I_*,
\end{aligned}
\qquad (133)
$$

where Ψ is a vertical length-scale, v is a characteristic vertical water velocity and υ is a mean sinking rate for phytoplankton. The scaled equation is

$$\frac{\partial B_*}{\partial t_*} + \frac{v}{\chi P_m^B}\left(\vec{u}_* \cdot \vec{\nabla}_H B_* + w\frac{\partial B_*}{\partial z_*}\right) - \frac{\upsilon w}{\chi P_m^B}\frac{\partial B_*}{\partial z_*}$$

$$= \frac{K_H}{\xi^2 P_m^B}\nabla^2 B_* + \frac{K_v}{\chi^2 P_m^B}\frac{\partial^2 B_*}{\partial z_*^2} + \left(\frac{I_*}{1 + I_*}\right)B_*. \quad (134)$$

Looking at the term corresponding to the one we discussed for the last example, we see that one of the criteria for similarity between a system and its microcosm is that the factor $(K_H/\{\xi^2 P^B\})$ is numerically equal in the two systems. The further complication is that P_m^B is now involved in the scaling of the light term. Again, the sinking rate of the phytoplankton cannot be modified unless we change the density of the water (ignoring the physiological causes for differential sinking rate), not a desirable step. But the vertical scale of the microcosm will certainly be less than that of the real system. The physiological constants α and P_m^B will presumably be equal in both systems. One could think of increasing the optical extinction coefficient in the microcosm to incorporate a light gradient in the microcosm similar to that in the real world, allowing for the reduced vertical scale of the microcosm. Here the difficulty is that in neither system will the extinction coefficient be independent of the biomass.

All of these considerations do not themselves invalidate completely the basic principle of microcosm research. But they should be sufficient to show that extreme caution is required if the scaling is to be done legitimately. It is necessary to consider what the effect of scaling one factor, such as the turbulence, will be on other factors, such as growth rate. Scales cannot be chosen independently.

The best method of approach would be to look at the scaled master equation for the real system (Equation (134) or analogue), decide what are the leading terms by comparing coefficients, and to build the microcosm with a view to making the leading coefficient close to its magnitude in the real world. At the same time what the features of the real world are that must be reproduced in the microcosm must be borne in mind. For example, for some studies it may be considered essential to duplicate vertical structure in the microcosm even if vertical effects do not dominate the master equation for the real world. This has to be taken into account when the scaling is done.

A fundamental limitation on all these studies in which biological processes are embedded in a dynamic physical system will be that both the physics and biology operate on the same time-scale, and in scale models we cannot moderate or adjust the biological rates to compensate for changes in physical rates that are a consequence of reducing the physical scales in the system. For this reason we expect microcosms to be more successful, in general, for organisms with shorter generation times.

A more subtle complication is related to the frequency of forcing the microcosm. For example, Steele *et al.* (1977) found that higher-frequency fluctuations in the physical fields of the real system were completely absent in the microcosm he studied (*in situ* plastic column). The theory required to examine these effects in detail has not been developed, although Roughgarden (1974) has shown how to calculate the influence of spatial inhomogeneities of various characteristic scales on biological rate equations.

3.3   Physical transports[1]

Marine life is strongly influenced by the hydro- and thermodynamic state of the sea, described by the field of motion, pressure, temperature, salinity and density, which influences not only the physiological processes of the individual organisms, but also the aggregation, stability and spreading of populations.

The motions in the sea are turbulent and extend over wide ranges in frequency/wave number space. Their importance for marine biological modelling lies in their function as the transport mechanism, not only for quantities

---

1. By G. Radach, J. M. Colebrook and M. J. Fasham.

like heat and nutrients, but also for the populations, providing an energy subsidy to the biological system (Mann, 1974; Margalef, 1978). By the formation of fronts and eddies, the hydrodynamic field may also provide a mechanism for partially isolating one population from another which may have important consequences for population dynamics and the development of spatial pattern (Wiebe *et al.*, 1976; Pingree, 1978).

Measurements of horizontal velocities can usually be used to calculate transports on the scales resolved by the sampling methods. But estimates of the often considerable transports, which are caused by the unresolved subgrid scale motions, can only be obtained by the use of physical assumptions for the parameterization of these motions.

In the very productive areas such as the mid-latitudes, where seasonal thermoclines develop, and upwelling areas, biological dynamics are largely controlled by physical transports which cannot yet be measured accurately, particularly those caused by vertical velocities. The interaction of these motions with sinking is a very important factor in the biology of phytoplankton (Margalef, 1978).

In mid-latitudes, it is the decrease of vertical turbulent transports, connected with the development of the thermal layering of the upper water masses, which determines the start of the spring plankton bloom (Sverdrup, 1953) and its intensity (Riley *et al.*, 1949; Radach and Maier-Reimer, 1975).

Expressions for the local vertical eddy diffusivity coefficient K_v have been derived on the basis of different theoretical arguments (Munk and Anderson, 1948; Sundaram and Rehm, 1973; Kullenberg, 1974; Mellor and Durbin, 1975). Values of this coefficient derived from measurements of the diurnal temperature cycle do not allow us to define a general vertical structure of K_v. The reason for this may be the fact that different physical processes (wind mixing, convection, current shear and wave breaking) contribute to the turbulent motions and that it is not clear how the local energy input from the atmosphere is distributed among the different processes (Pollard, 1977).

It is becoming apparent that very weak or even transient vertical stability may be sufficient to initiate a phytoplankton bloom. Nevertheless the physical upper-layer models, still in the developmental phase with respect to some observed features, could provide a useful method for determining the turbulent transports in time and space for biological models. The one-dimensional models are able to simulate the diurnal and seasonal temperature variations in the vertical dimension quite realistically, for regions where horizontal advection can be neglected. Many of the one-dimensional upper-layer models show weaknesses in reproducing fast changes such as occur after storms (Niiler and Kraus, 1977), but for many biological problems connected with the seasonal thermocline they seem sufficiently accurate to make use of them.

For modelling the biological dynamics in upwelling areas the crucial vertical velocity w is required, but it cannot be measured accurately enough. Usually, therefore, the continuity equation is used to calculate w from horizontal velocities, either directly from measurements, or using a physical upwelling

122

model. The physical dynamics may be calculated either decoupled from biological dynamics (e.g. Walsh, 1975) or they may be coupled to the biological dynamics and calculated simultaneously (e.g. Wroblewski, 1977). The latter procedure seems advisable for studying the sensitivity of the biological system to physical environmental changes.

During recent years attempts have been made to describe the spatial variability of chlorophyll using spectral analysis and to interpret these spectra using the theory of turbulence (Denman and Platt, 1976; Denman et al., 1977). Physical parameters of interest in this study are the Richardson number Ri and the rate of viscous dissipation of turbulent kinetic energy, ε. Denman and Mackas (1978) have reviewed recent instrumental developments which are potentially capable of mapping these parameters down to scales of 1 m.

In the mesoscale range (i.e. time-scales ~ 100 days, length-scales ~ 100 km) the physical processes of most interest to biologists are probably the structure and dynamics of rings and eddies. Rings are rotating parcels of water that have been 'spawned' from a strong current system and usually have discernible water-mass properties that distinguish them from their surroundings. An example is the cold core rings found to the east of the Gulf Stream (Lai and Richardson, 1977) which appear to support distinctive faunal assemblages (Wiebe et al., 1976). The term eddy is usually reserved for features that have no obvious anomalous water-mass properties but are cyclonic or anticyclonic undulations of the mid-ocean isopycnal surfaces. The physical properties of these eddies have been extensively studied during the Mode series of experiments (Mode Group, 1978) but their biological characteristics are as yet unknown. The theoretical modelling of rings (Flierl, 1977) and eddies (Mode Group, 1978) is proceeding apace and will probably be well developed by the time biological oceanographers attempt the formidable task of modelling deep-ocean pelagic ecosystems.

The view of an ecosystem derived from a time-series of, say, several decades differs markedly from that derived from observations covering a single seasonal cycle. And yet, while one of the main objectives of the development of production models is the prediction of year-to-year changes, the contemporary approaches to modelling are strongly influenced by views of the ecosystem based on the seasonal cycle.

There are very few time-series relating to marine organisms, other than fisheries data, covering periods of more than a few years. The Continuous Plankton Recorder survey, the CalCOFI survey, the observations at OWS Papa (North Pacific) and station E (English Channel) are among the very few approaching a decade or more.

Thus, the available interpretations of long-term changes are confined to these few series. They are limited in scope and location. The general impression is, however, that physical transport and primarily advection plays a key role. This is certainly true for the North Atlantic (Colebrook, 1978), for the California Current (Wickett, 1967; Colebrook, 1977) and for the English Channel (Russel et al., 1971).

In the absence of direct data for advection these interpretations are based on indirect assessments derived from surface temperatures, sea levels or atmospheric pressures. This severely limits studies of the mechanisms by which variation in advection produces the biological changes. There is clearly a need for more extensive observations of variations in advection on time-scales of years to decades and for the development of physical models of the relevant systems.

The emphasis on advection is a symptom of the growing realization of the role of changes in climate and climate-related phenomena in determining long-term changes in marine ecosystems. Such changes relate to large areas and long time-periods. At some stage, it will be necessary for biological production models to converge with models of ocean circulation and with even larger-scale climatological models in order to contribute to problems such as the global increase in carbon dioxide resulting from the consumption of fossil fuels. And, to quote Unesco (1974a), 'sponsoring bodies should give every practicable encouragement to individuals and to agencies, world-wide, in assembling time-series data from the ocean to further our understanding of the actual processes of variability...'.

3.4 Statistical design of field programmes

3.4.1 THE IMPORTANCE OF SCALE[1]

In the marine habitat populations exhibit great variability as a result of the interactions of biological and physical processes. Investigation of this variability is of great importance for many biological questions: for judgement about the representativeness of a single measurement, for the design of field programmes and for the description and explanation of the dynamics responsible for the observed variability.

Assuming that there are biological hypotheses or models at the start of a field programme, we wish to ensure that the results will validate or falsify the hypothesis and that progress will be made in solving the biological problems. The difficult questions that arise are: what has to be measured, and what is the experimental design that will ensure a decision about the validity of the hypothesis?

First, we must know the scales in space and time which dominate the problem to be investigated, i.e. the scales on which the biological, chemical and physical processes interact. In other words, knowledge of the variability of the variables or parameters under consideration is the basis of a sensibly designed field programme. To complicate matters, biological and physical scales of variability may be different. If the scales of a process are not known, experiments to explore these scales should be planned on the basis of existing knowledge.

1. By G. Radach and K. H. Mann.

Although much effort in the last few years has been directed towards augmenting our meagre knowledge of the variability of biological quantities in space and time, there are still gaps over wide ranges. The reason for this is to be found in the methods of biological oceanography. In physical oceanography automatically recording instruments such as current meters, thermistors and tide gauges have been in use for a long time, but in biological oceanography automatically recording instruments are available for only a few parameters, and those are comparatively recent.

Continuous recording, preferably with constant time increments, is a prerequisite for the use of such powerful mathematical tools as time-series analysis. This in turn has made possible, in physical oceanography, the representation of the energy distribution of water movements in frequency/wave number space. Analogous developments have taken place in biological oceanography in the last ten years. Continuous measurements of chlorophyll, nitrate and zooplankton have been related to temperature and given insights into the scale of horizontal patchiness (Platt, 1972; Denman, 1976; Fasham and Pugh, 1976; Mackas, 1977; Richerson et al., 1978; Denman and Mackas, 1978; Steele and Henderson, 1978). The data have been used to develop theories of phytoplankton patchiness (Evans et al., 1977; Wroblewski, 1977; Okubo, 1978). Measurements and theories show that the dominant length-scale of a phytoplankton patch is a few kilometres.

Attempts to represent biological quantities in frequency/wave number space have led to a diagram for the variability of the biomass zooplankton (Haury et al., 1978; see Fig. 13). The greatest difficulty in such an undertaking lies in the fact that biological quantities such as zooplankton biomass do not obey a conservation law, and are not transported through the scales in a manner analogous to turbulent energy in physical oceanography. Nor is there a unique measure of variability of biomass that can be attached to the vertical axis (in the figure). On the other hand, diagrams like this one can serve to sort out the complex biological mixtures of scales for different quantities, and thus show lacunae in our knowledge.

While the approach of Haury et al. (1978) aims at the variability of a special group of organisms, other studies have tried to compare physiologically determined characteristics of different groups of organisms via space- and time-scales. The spectrum of Sheldon et al. (1972), relating size classes to growth rates, is a good example (Fig. 14).

Following Haury et al. (1978) several space- and time-scales can be distinguished:

Space-scales		*Time-scales*	
Mega scale	> 3,000 km	Climatic scale	> few years
Macro scale	1,000–3,000 km	Meso scale	few months/few years
Meso scale	100–1,000 km	Seasonal scale	1 year
Coarse scale	1–100 km	*Weather-scale*	few days/1 year
Fine scale	1–1,000 m		
Micro scale	1–100 cm	*Diurnal-scale*	1 day

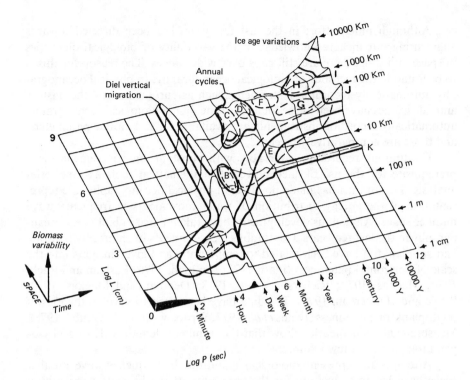

Figure 13
A semi-quantitative three-dimensional representation of relative variability in zooplankton biomass over a range of time- and space-scales (from Haury *et al.*, 1978). A, 'micro' patches; B, swarms; C, upwelling; D, eddies and rings; E, island effects; F, 'el niño'-type events; G, small ocean basins; H, biogeographic provinces; I, currents and oceanic fronts: length; J, currents: width; K, oceanic fronts: width

Of all the scales the best known in marine biology are the spatial coarse scale and the diurnal-to-seasonal time-scales.

To resolve the scales by measurements, certain minimum requirements have to be fulfilled. The sampling density must be sufficiently small that, e.g., the time increment, Δt, between two samples, determining the Nyquist frequency $w = \pi/\Delta t$ (which gives the smallest resolvable time-scale) is sufficiently small that a good resolution of the frequencies investigated is ensured.

If the sampling density should make possible the resolution of finer scales than necessary for the problem under consideration, no difficulties will arise as long as the relevant scales are resolved, too. The effect of processes of finer scale can be eliminated by filtering techniques. On the other hand, if the sampling density does not allow the resolution of the scales of the problem, the measurements will be of little value. This means that the problem of matching the space-time scales of the model to those of the physical and biological systems under consideration must be solved by the experimental design.

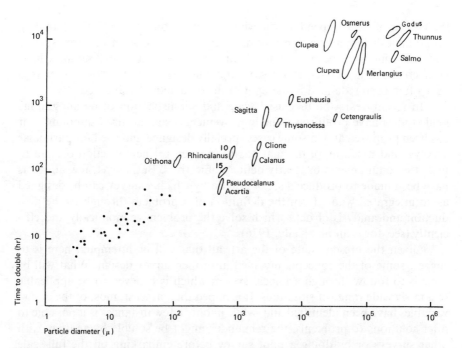

Figure 14
The relationship between rate of production and particle size. The numbers near to the
Rhincalanus patches indicate the temperature at which the growth took place. The uppermost
of the two *Clupea* areas represents *C. sprattus*. The lower area represents both the Atlantic
(*C. harengus*) and Pacific (*C. pallasi*) herring (from Sheldon *et al.*, 1972)

3.4.2 SURVEY DESIGN[1]

For the purposes of this account, a survey will be defined as a structured set
of field observations resulting in the production of data relating to marine
organisms. There is an enormous range of such surveys and it is obviously
impossible to consider them all in detail. This account will be restricted to a
consideration of general principles and their application to the limited field of
plankton surveys. That there is a problem in designing surveys stems largely
from the fact that the concentrations of organisms are extremely variable over
wide ranges of time- and space-scales as outlined in the previous section.

No one sampling method and no one sampling programme can conceivably
provide information about the variability for more than relatively narrow
zones of the time and space ranges. Variability outside the selected zones
interferes with this information. A major part of survey design is concerned
with minimizing the effects of this interference.

Surveys tend to be designed empirically, the designs depending on the
experience of the designer. It is, perhaps, not too great an exaggeration to say
that in the past many plankton surveys have been put together fortuitously,

1. By J. M. Colebrook and M. J. Fasham.

the choice of gear depending on what has been on the shelf, the sampling programme depending on ship-time made available during other cruises, sorting and counting have been over-ambitious, with the result that samples have been left standing on the shelf for years and finally the objectives of such surveys have often been redefined in the light of the data actually obtained.

In recent years the increased scale and sophistication of oceanographic field studies have produced some improvements; standardized sampling gear has been proposed and in some cases specially designed and used for particular surveys, and much more thought has been given to the collection of data to meet the requirements of clearly defined objectives. But, as yet, few attempts have been made to produce a set of criteria by which a survey can be designed as an integrated whole from the definition of a problem through to the production and analysis of data which solve the problem economically and efficiently (see, for example, Kelly, 1976).

Given the present state of the art, all that will be attempted here is to suggest some of the concepts involved in proper survey design. What will be done is to follow through a logical system which is believed to be applicable to a fairly wide range of situations. It is hoped that, at least, most of the design variables have been identified and, while in only a few instances is it possible to offer solutions to problems, the remainder might be soluble by analogy with other surveys or by doing a pilot survey before embarking on the full-scale operation.

The design problem is divisible into three main stages:
1. The definition of the objectives of the survey.
2. Strategic considerations—the amplification of the objectives in terms of specific quantitative variables required from the results of sampling.
3. Tactical considerations—the formulation of field procedures to meet these requirements.

Survey objectives

The selection of objectives is not strictly part of the design process. It must be stressed, however, that the formulation of objectives is a crucial phase. The objectives should be defined as precisely as possible and in a form capable of realization and amplification at the strategic planning phase.

The strategic plan

It is necessary, at this stage, to consider what information is required to achieve the objectives of the survey. This information is in the form of, or is derived from, quantitative variables—the result variables—containing 'signals' observed against a background of 'noise'. The result variables may be the counts from samples. They are more likely, however, to be data derived from the results of sampling by averaging or transformation or the results of, say, multivariate or time-series analyses.

The strategic planning phase is primarily concerned with the form of the result variables; it is necessary to decide what variables should be measured, the co-ordinates of the variables in space and time, the resolution along the co-ordinator and the derived precision of the variables. If the result variable is derived from an analytical process, say a principal component or a power spectrum, then obviously these factors have to be considered in terms of the data which are fed into the analysis in order to achieve specifications or result variables that can be implemented at the tactical level.

The type of information

In general terms this should be implicit in the definition of objectives. Where there is a choice, the problem at the strategic stage lies in assessing the required level of discrimination. In, for example, a survey involving the collection of samples of organisms, is it necessary to perform identification to the species, or would discrimination to, say, a few size categories be sufficient? A high level of discrimination provides flexibility in the subsequent design. On the other hand, aggregation of taxa may provide opportunities for simplification of field or laboratory procedures permitting the processing of a larger number of samples.

A pilot survey may provide some of the answers, or alternatively the results from an existing survey can be examined, and an attempt made to simulate the kinds of observations that are under consideration.

The format of the data

Having decided on the types of information to be collected it is necessary to consider the statistical or other analytical procedures involved in deriving the final result variables. These will influence the spatial and temporal patterns of the field observations. Detailed sampling patterns will be discussed at the tactical level; here the prime consideration is the selection of co-ordinates in terms of which the data are to be expressed at the stage immediately prior to final analysis.

A simple structure is a table with each column representing a station or a set of depths at a single station and each row representing a species or other taxonomic or quasi-taxonomic (e.g. size category) entity.

Such structures can be elaborated by generating a series of tables for specific time intervals, months or years for example.

Three- or four-dimensional data sets are capable of expression in a variety of two-dimensional subsets which are amenable to analysis by a variety of methods such as analysis of variance, multiple regression, principal components analysis etc. Such methods can provide information on the validity of the data, based on its internal structure, and can also provide expressions of interrelationships and coherent elements of variability.

Most sample designs imply a fairly rigid data structure with observations available for regular times and uniform spatial patterns. Occasional missing values can be accommodated either by interpolation or by suitable procedures in analysis, but irregular spatial and temporal patterns are difficult to deal with without a considerable loss of information.

Maximizing information

This is a function of the signal-to-noise ratio and is related to both resolution in space and time and to precision of estimation. This is obviously a key factor in the survey design and is one of the most difficult to assess.

Given a properly defined objective it is not difficult to produce a specification for the result variables that will satisfy the requirements. But, in the absence of information about the variability of the population to be sampled in relation to relevant time- and space-scales, the formulation of a specification that can be implemented is impossible.

Any assessment of quantity of information presupposes some knowledge of the variability of the population to be sampled. If such information cannot be obtained from published data then it will be necessary to carry out a pilot survey. In relation to spatial heterogeneity such a survey can provide the required information. There is an obvious problem, however, in producing information about time-varying parameters, particularly if the time element is longer than a year. There is as yet no complete substitute, in the planning phase, for intuition based on experience.

This makes it desirable to be able to produce a posteriori estimates of precision. Provision for this should be built into the survey design. Cassie (1971) considers the use of stratified random-sampling patterns in the estimation of errors based on linear statistical models. Its application is limited to station-based sampling techniques and it implies that an appreciable proportion of the sampling effort is devoted solely to the estimation of errors. Colebrook (1969) considers the use of multivariate-analysis techniques in the search for coherence and pattern in sets of comparable variables. The calculation of precise significance levels is difficult and the assessment of the reality of the derived patterns of variability has to be based largely on intention and experience. On the other hand, there is minimal redundancy of sampling effort, although it may require analysis of samples at a lower level (say, species as opposed to total biomass) than may be required by the objectives of the survey. Colebrook (1975) used simulated replication in a study of some of the result variables obtained from the Continuous Plankton Recorder survey (Glover, 1967). A negative binomial distribution was used to represent spatial heterogeneity, with parameters based on the observations. This method is applicable to most surveys but is dependent on the suitability of the model chosen to represent the distribution of the population as sampled.

The problem of the assessment of quantity of information is the major limiting factor in survey design. The reliance of a posteriori estimation of precision may lead to the necessity of redefining objectives in the light of what can actually be achieved, and the designer is left with little alternative but to go for the maximum precision permitted by resource limitations. Considerations of efficiency have to take second place in the design.

Since populations are usually aggregated, precision of estimation is primarily a function of the number of samples collected as opposed to the number of organisms counted (see below). Precision is achieved by averaging the results of sets of samples and the designer may have to sacrifice resolution in space or time for precision of individual estimates.

The tactical plan

This stage is concerned with the implementation of the strategic plan into practical field and laboratory procedures. The main aspects are:
1. Sampling pattern and density in space.
2. Sampling pattern and density in time.
3. Sampling gear.
4. Sample size and form.
5. Counting or measurement procedures.
6. Data processing and analysis.

The main task of the tactical plan is to arrive at the best compromise between the requirement established at the strategic phase and the inevitable resource limitations associated with gear, deployment, manpower and data-processing capacity. These features are basically unique to each survey but some generalizations are possible.

The mechanics of sampling, subsampling, counting and data processing are adequately covered in available literature. Unesco (1968, 1974b), Edmondson and Winberg (1971), Holme and McIntyre (1971), Sournia (1978) and Pugh (1978) are all useful sources of information.

With respect to field procedures the key features are the sampling patterns in space and time. In most situations the requirements will be such that resources will be a prime limiting factor. The selection of gear will be determined by the need to collect a sufficient number of samples and the counting and measurement procedures will be influenced by the need to process the samples in a reasonable time. In determining sampling pattern we are immediately faced with the same difficulties as were considered in the assessment of quantity of information in the result variables. In this instance the quantity of information is directly related to the number and pattern of samples in space and time and since populations are generally aggregated and also vary in time some knowledge of these aspects of variability is an essential prerequisite to establishing patterns of sampling.

Computer simulation of the population to be sampled probably offers the most general method of studying sampling problems. There is a wide range of methods that might be used but this discussion will be limited to one simple model that illustrates the main principles involved.

The negative binomial distribution has been used to parameterize small-

131

scale spatial heterogeneity (see, for example, Colebrook, 1975, and Taylor, 1953). Its general term is:

$$P_c = \frac{R^c(k + c - 1)!}{q^k c!(k - 1)}. \tag{135}$$

where $R = P/q$ and $p = 1 - q$. The parameters can be estimated from the mean and variance of observations from, say, a pilot survey, thus

$$k = \bar{x}^2/(\bar{x} - s^2)$$

and

$$q = s^2/\bar{x}. \tag{136}$$

A complete series is most easily calculated using

$$P_o = 1/q^k; \quad \text{and} \quad P_c = P_{c-1} \cdot R(k + c - 1)/c; \qquad c = 1, 2, 3 \ldots, \tag{137}$$

and 'replicate' samples can be selected from the series using random numbers. Colebrook (1975) used the negative binomial distribution to simulate spatial heterogeneity on the scale of sampling with the Continuous Plankton Recorder. The exponent R was assumed to be constant and was estimated from observations. Using time-series of monthly means, also derived from observations, calculated frequency distributions were used to provide simulated replicate sampling in trial-and-error assessments of various sampling and counting schemes. Other stochastic models of spatial heterogeneity are described by Fasham (1978a).

The simulation of time-varying processes clearly presents considerable difficulties. Smith (1978) considers the problem of estimating the mean of time-varying parameters based on the properties of stationary stochastic processes, and some other models that could find application in this content are described in Section 2.2.5.

The general relationship between precision of estimation, the number of samples and the mean number of organisms per sample is illustrated in Smith's Figure 1, in which the standard error of the mean divided by the mean is plotted against the number of samples for means of 1, 10 and 100 and for values of the exponent of a negative binomial distribution ranging from 100, which is virtually equivalent to a Poisson distribution, to 0·5 representing a highly skewed distribution. A transition can clearly be seen from a situation in which precision is determined primarily by the number of organisms counted ($k = 100$) to one where precision is dominated by the number of samples collected ($k = 0·5$). The relatively slight improvement in precision at low values of k for means of between 10 and 100 should be noted in relation to decisions on sample size and how many organisms to count.

In Smith's Figure 2 are given probability distributions for means of 2, 10 and 50 for negative binomials with values of k of 0·5, 1, 2 and 100. These distributions in combination with the information on precision suggest an optimum sample size giving means of about 10. Very little is gained in terms of precision with larger averages and individual sample counts larger than about 60 should occur infrequently.

Cassie (1968) reaches much the same conclusions, based on similar reasoning. To quote Cassie, 'This phenomenon, which was probably first presented in the marine biological literature by Taylor (1953) in connection with demersal trawl sampling, has caused some controversy and is probably still suspect by many.' This is probably still true ten years later, which is why the argument has been repeated here.

In establishing counting procedures, consideration should be given to the use of counting categories (Colebrook, 1960). It is common practice to apply a logarithmic transformation to counts prior to processing, in which case the resolution of the counting procedures could be matched to the logarithmic scale. In general, averages of sets of samples from skewed distributions are influenced more by the relatively high proportion of low counts than by the low proportion of high counts, and some loss of resolution has very little effect.

3.5 Ecosystems under stress[1]

Introduction

Ecosystems, complex entities that they are, undergo manifold changes in response to external perturbations, such as rapid temperature change, sudden nutrient addition, osmotic pulses, etc. There appears to exist no clear consensus as to how to quantify the effects of stress under all circumstances (Thorp and Gibbons, 1978).

Nevertheless, any rational discussion of the topic of systems response to stress must proceed under an assumption as to how to quantify system change. Now, throughout this book we have chosen to place heavy emphasis upon the measurement of flows in ecosystems. In fact, in Section 2.2.2 the hypothesis was considered that the optimization of a state variable (the community ascendancy) synthesized from the network of flows provides the criterion for ecosystems development.

Ascendancy was taken to be the product of two trends: growth and development. The development, or integrity, of a network is a measure of how well the flows within the system relate one to another. If Q_i is the fraction of the aggregate network flow passing through compartment i, and f_{ij} is the fraction of throughput i which flows directly to sustain throughput j, then the average mutual sustenance of the system can be measured as

$$D = \sum_j \sum_k f_{kj} Q_k \ln\left(f_{kj} \Big/ \left[\sum_i f_{ij} Q_i \right] \right). \tag{138}$$

Other things being equal, this index is enhanced by increases in cycling, positive feedback and specialization of contributions—in short, those elements which characterize a 'developed' community (Margalef, 1968; Odum, 1969).

1. By R. E. Ulanowicz and K. H. Mann, Contribution No. 969 of the University of Maryland, Centre for Environmental and Estuarine Studies.

133

As a working postulate, we choose to describe the effects of stress upon a community in terms of the change induced in the development index, D. It should be made clear that this measure is intrinsic to the system and is not meant to convey a sense of benefit to man. For example, less-developed systems can be more productive (Margalef, 1968), and the upwelling and estuarine zones of the marine habitat are 'arrested' communities which yield a greater abundance, in a direct sense, to man than the more developed open-sea systems.

The usual perception is that external perturbations act to decrease ecosystem development. This is almost always the outcome of high-intensity stochastic stress. The response to lower-level or more predictable stresses, however, can be qualitatively quite different. Atlan (1974), for example, resolves the apparent paradox that low levels of stress are necessary for a system to undergo self-organization. In the biological realm, then, stress can sometimes give rise to more organized configurations.

Although the development index defined here gives some semblance of rigour to the discussion of community response to stress, it has the disadvantage that the data required to calculate the state variable are extremely costly to obtain. The ecosystem manager, like the engineer, often wishes to make a sound decision based on the quickest, most inexpensive data collection possible. The only requirement is that the variable computed from measurements should correlate well with the development index. In contrast, the ecosystems researcher is more concerned with the dynamic character of the community and would require data enabling a closer approximation to the actual response.

Simple correlates of development

Perhaps the simplest indicators of ecosystem stage are ratios of extensive system variables. For example, more developed communities appear to support more biomass, B, for each unit of energy throughput, E. The ratio B/E might, therefore, serve as an approximation to community development (Margalef, 1968). Likewise, the ratio of standing crop to gross productivity increases in relatively undisturbed systems. Odum (1969) lists twenty-two other similar ecosystem attributes which appear to correlate with increasing community development.

Well-developed pelagic communities have roughly the same volume concentration of various-sized particles (Sheldon et al., 1972). Stressed communities tend to become depauperate in larger-sized organisms (Kerr, 1974b). The relative ease with which particle size distributions may be automatically measured makes this assay most attractive (see also Section 2.2.3).

Probably the most widely measured index of system response to stress is species diversity. It is likely that species diversity reflects underlying flow diversity, which in turn is necessary (but not sufficient) to allow greater network development. Species diversities are thus seen to be indicators of the upper bound to ecosystem community development. Any significant decline in the limits to development is likely to impact upon development itself.

134

Sanders (1968), for example, showed clearly that benthic communities which had been subjected to physically stable conditions for long periods of time had greater numbers of species than communities in environments which were physically unstable, particularly if the physical variability was irregular and unpredictable. Johnson (1970) extended Sanders's ideas and showed that stresses such as storms, salinity changes, or sudden temperature changes cause regression of benthic communities towards earlier stages of development. Conversely, periods of environmental stability led to further community development with higher species numbers. In 1978 Sanders showed that oil pollution was a stress that produced changes analogous to those seen under the influence of natural stresses, i.e. reduction in species diversity and population instability.

A note of caution must be sounded here. The theoretical development index is a property of the whole ecosystem, and it might be expected that a species diversity index would correlate well with the development index if it, too, related to the whole biological community in the system. In practice it has been possible to derive species diversity indices only for rather restricted subsets of a biological community, e.g. a few groups of benthic invertebrates, or the zooplankton retained in a standard net. Diversity indices representing the whole community from bacteria to vertebrates have not been obtained and would be almost impossible in practice, since the scales of spatial distribution of the various components of the biota are so different. Hence, we must remember that the literature on diversity indices refers only to small subsystems, which may or may not behave in the same way as the total system.

However, species diversities calculated from net tows or bottom counts and size diversities derived from particle counts (Parsons, 1969) are readily obtainable and may well be important correlates to the theoretical development index. Spatial, or pattern, diversity is rarely mentioned in connection with stress, but should also play an analogous role in assessing environmental impact.

Dynamic indicators of a stressed community

The impingement of a stress upon a community and the subsequent system response are both dynamic processes. It is unlikely that any of the aforementioned snapshot indices are likely to be the sole descriptors of time-dependent behaviour. The information necessary to address the more philosophical question of *how* ecosystems cope with stress is sure to be more extensive and costly to obtain. It may be that the satisfactory exposition of ecosystem stress response will take the form of a mathematical model (e.g. Odum, 1971; Austin and Cook, 1974), but it appears more likely that an explanation at the systems level will require the use of variables which characterize the composite community.

The reader's attention, therefore, is directed to the second section of this book on holistic methods in ecology. There the point was emphasized over and again that answers to questions of community dynamics will most probably

135

result from data on ecosystems processes. Thus, the measurement of the eco-systems flows should be given research priority over the measurement of compartmental contents. In particular, just knowing the weighted digraph of community exchange rates will allow for a number of alternative assessments of the response to stress.

Odum, for example, indicated that cycling accounts for a greater percentage of total system flow in more developed communities. Finn (1976) showed how input–output analysis can be used to derive a cycling index for the entire assemblage. Hypothetically, excessive stress should cause a decrease in Finn's cycling index. In principle, cycled flows can be aggregated according to the trophic length of their loops. In the absence of inordinate stress, cyclic controls develop involving loops of greater length. Control response time diminishes (Golley, 1974). The effect of inordinate stress would presumably be to diminish the magnitude of higher-order cycles. Trophic flow spectra of impacted systems should, therefore, show diminished magnitude of cycling at all levels and particular decline in the higher-order control loops.

Kerr (1974b) has remarked that stressed pelagic communities tend to possess a smaller proportion of larger-sized organisms. In pelagic systems larger-sized animals also tend to feed higher on the trophic chain. One would expect, then, that stress would reduce the ranks of organisms feeding at higher levels. Ulanowicz and Kemp (1979) present an algorithm for apportioning populations into trophic aggregates. Homer and Kemp (1977) provide data from a thermally stressed marsh community showing that the lower trophic aggregations remain practically unperturbed, whereas the higher trophic compartments are drastically impacted. Interestingly, the species list in the two communities did not differ significantly. Many animals normally feeding high in the trophic chain were subsisting on lower forms, and their trophic index (Levine, 1979) had fallen.

If a full description of all ecosystems flows were available, then the development index, D, could be directly calculated and its response to various stresses observed. In practice this ensemble of data is formidable and almost impossible to collect in its entirety. An alternative to knowing all the flows is a cross-correlation of all the changes of the various compartments of the ecosystem. In the parlance of information theory, each compartment would constitute a channel and the *changes* in compartmental content (of mass or energy) would comprise a signal within that channel. The average mutual information among the various channels should reflect the intrasystem control in a manner similar to Mulholland's (1975) index for flow structures. Unlike the flow analysis, however, the residual uncertainty cannot be attributed to system overhead. This does not detract from the utility of the information measure as an index of system response to stress, i.e. average mutual information would decline under damaging perturbation. The subsection on particle-size spectra describes how measurement of particle size might be employed to estimate the dynamic information of a pelagic ecosystem.

Characterization of stress

This section would be incomplete without a few words on characterizing the causal perturbations. Exogenous stresses acting on a system are described by their magnitude, duration and physical extent. While stresses are often conveniently approximated by step functions or impulse functions, other perturbations are temporally and spatially more complex. More often than not, disturbances are stochastic in time and space, and such functions are often described as temporal and spatial spectra (Fasham, 1978*b*; Platt and Denman, 1975). Spectral descriptions of ecosystems are growing in popularity. In addition to the temporal and spatial classifications of ecosystem changes, particle size spectra and 'trophic' spectra (Ulanowicz and Kemp, 1979) are finding use in holistic description of systems.

The implications of the trend towards spectral description of ecosystems are straightforward. The dynamics of community behaviour can be effectively summarized by spectra. One strives to describe stress and systems response in the same general terms. Therefore, the chance discovery of a stressed community by any of the previously described methods is an impetus to describe *both* the system dynamics and the abiotic forcing functions in terms of appropriate spectra. Comparison of the two sets of data is very likely to result in a more enlightened view of the dynamics of stress upon ecosystems.

4

General conclusions[1]

It is important to emphasize that ecological models will never be the universal panacea to biological oceanography; nor will they ever replace the need for well-designed observations on the real world. But if they are used intelligently, good models should help make the observational process more economical, more penetrating and the interpretation of the results more stimulating. Misused, models of any kind will lead to, at best, self-deception on the part of the user, and at worst the waste of money and manpower in the pursuit of a poorly conceived sampling programme with the possibility of subsequent provision of bad advice to the managers of an important resource.

The various approaches to modelling that we have considered might be split into two broad classes: reductionist and holistic. Reductionist models are more commonly used and more readily understood by biologists without a strong mathematical background; they offer the advantage that, given sufficient resources, each component model can be researched in detail. Their principal disadvantage is that the results are usually location-specific. (Note that this may be an advantage in a specific management application.) We have learned already from reductionist methods a great deal about how subsystems interact, but the principles of whole-ecosystem dynamics still elude us.

Holistic approaches to whole-ecosystem dynamics, on the other hand, are relatively unexplored so far. The analyses are not usually location-specific and hold the promise that we may still learn something new about the general principles of ecosystem organization.

We have been impressed again and again in the course of our review of available models, whether holistic or reductionist, by the paramount importance of measuring physiological rates for the computation of ecological fluxes. Observational effort in the past has emphasized biomasses at the expense of fluxes, even though it must be clear that both kinds of measurements deserve equal status. Quantification of rates is the *only* avenue by which models can be driven from the static to the dynamic. A static representation of a dynamic system is of but limited utility.

We have also been impressed with the value of keeping constantly in mind

---

1. By T. Platt.

138

the idea of scales in time and space. Adequate consideration of scale will help avoid a mismatch between sampling intervals and desired resolution of the data. It should be clear that all processes with characteristic frequency greater than the inverse of the fundamental time-scale of the model will necessarily be parameterized. Processes with lower characteristic frequency may be treated deterministically. Implicit in the choice of time-scale for a model is the decision that processes on a smaller time-scale will be time-averaged. For example, if the fundamental time-step in a dynamic simulation model is one day, then the detailed, non-linear effects of, say, vertical migration of zooplankton feeding will be averaged out. This is not necessarily a bad thing, but the modeller should be aware of it. Many models of phytoplankton–zooplankton interaction when applied to pieces of the real ocean show that zooplankton do not have at their command sufficient resources of phytoplankton to satisfy their basal metabolic needs, if the averaging scales of the model are too great. Proper consideration of the scales of horizontal and vertical variability in the abundance of phyto-plankton shows that, locally, zooplankton can, in fact, find enough food to grow and reproduce, a phenomenon we know already must exist from our observation of the real world.

Let us be clear that mathematical modelling is just another tool that can be helpful in the right hands. Let us also be equally clear that its misuse is just as shameful as the misapplication of any other kind of analysis, whether it be chemical, physical or whatever. Used well it can save in time, money and effort. Used badly it will lead only to waste.

References

ALLDREDGE, A. 1976. Appendicularians. *Sci. Am.*, vol. 235, p. 94–102.

ANKAR, S. 1977. The soft bottom ecosystem of the northern Baltic proper with special reference to the macrofauna. *Contrib. Askoe Lab. Univ. Stockh.*, no. 19. 62 p.

APHA. 1971. *Standard methods for the examination of water and waste water*, vol. 13, p. 330–33. Washington, D.C., American Public Health Association.

ARMSTRONG, F. A. J.; WILLIAMS, P. M.; STRICKLAND, J. D. H. 1966. Photo-oxidation of organic matter in sea water by ultra-violet radiation. *Nature* (Lond.), vol. 211, p. 481–3.

ATLAN, H. 1974. On a formal definition of organization. *J. Theor. Biol.*, vol. 45, p. 295–304.

AUSTIN, M. P.; COOK, B. G. 1974. Ecosystem stability: a result from an abstract simulation. *J. Theor. Biol.*, vol. 45, p. 435–58.

AZAM, F.; HOLM-HANSEN, O. 1973. Use of tritiated substrates in the study of heterotrophy in seawater. *Mar. Biol.*, vol. 23, p. 191–6.

BADER, H. 1970. The hyperbolic distribution of particle sizes. *J. Geophys. Res.*, vol. 75, p. 2823–30.

BAKER, J. H.; BRADNAM, L. A. 1976. The role of bacteria in the nutrition of aquatic detritivores. *Oecologia*, vol. 24, p. 95–104.

BANNISTER, T. 1979. Quantitative description of steady state, nutrient-saturated algal growth, including adaptation. *Limnol. Oceanogr.*, vol. 24(1), p. 76–96.

BANSE, K. 1974. On the role of bacterioplankton in the tropical ocean. *Mar. Biol.*, vol. 24, p. 1–5.

——. 1977. Determining the carbon-to-chlorophyll ratio of natural phytoplankton. *Mar. Biol.*, vol. 41, p. 199–212.

BARSDATE, R. J.; DUGDALE, R. C. 1965. Rapid conversion of organic nitrogen to N_2 for mass spectrometry: an automated Dumas procedure. *Anal. Biochem.*, vol. 13, p. 1–5.

BARTLETT, M. S. 1966. *An introduction to stochastic processes*. Cambridge, Cambridge University Press. 362 p.

——. 1975. Comment on paper by Cliff, A. D. and Ord, J. K. *J. Roy. Statist. Soc.*, vol. 37, no. 3, p. 297–347.

BERMAN, T. 1973. Modifications in filtration methods for the measurement of inorganic ^{14}C uptake by photosynthesizing algae. *J. Phycol.*, vol. 9, p. 327–30.

BLACKMAN, F. F. 1905. Optima and limiting factors. *Ann. Bot.*, vol. 19, p. 281–95.

BLISS, C. I. 1965. An analysis of some insect trap records. In: G. P. Patil (ed.), *Classical and contagious discrete distributions*, p. 385. Calcutta, Statistical Publishing Society.

BOHLOOL, B. B.; WIEBE, W. J. 1978. Nitrogen-fixing communities in an intertidal ecosystem. *Can. J. Microbiol.*, vol. 24, p. 932–8.

BOVER, D. C. C. 1978. Moment equation methods for nonlinear stochastic systems. *J. Math. Anal. Appl.*, vol. 65, p. 306–20.

BRAY, R. N.; HIXON, M. A. 1978. Night-Shocker: predatory behaviour of Pacific electric ray (*Torpedo californica*). *Science* (Wash.), vol. 200, p. 333–4.

BREGNBALLE, F. 1961. Plaice and flounder as consumers of the microscopic bottom fauna. *Medd. Dan. Fisk.-Havunders.*, n.s. vol. 3, p. 133–82.

BRIAND, F.; MCCAULEY, E. 1978. Cybernetic mechanisms in lake plankton systems: how to control undesirable algae *Nature* (Lond.), vol. 273, p. 228–30.

BROOKS, F. R.; DIMBAT, M.; TRESEDOR, R. S.; LYKKEN, L. L. 1952. Determination of small amounts of molecular oxygen in gases and liquids. *Anal. Chem.*, vol. 24, p. 520–24.

BRYLINSKY, M. 1972. Steady-state sensitivity analysis of energy flow in a marine system. In: B. C. Patten (ed.), *Systems analysis and simulation in ecology*, vol. II, p. 81–101. New York, N.Y., Academic Press.

BRYLINSKY, M; MANN, K. H. 1973. An analysis of factors governing productivity in lakes and reservoirs. *Limnol. Oceanogr.*, vol. 18, p. 1–4.

BULMER, M. G. 1974. A statistical analysis of the 10-year cycle in Canada. *J. Anim. Ecol.*, vol. 43, p. 701–18.

——. 1976. The theory of predator–prey oscillations. *Theor. Popul. Biol.*, vol. 9, p. 137–50.

CAMPBELL, M. J.; WALKER, A. M. 1977. A survey of statistical work on the MacKenzie River series of annual Canadian lynx trappings for the years 1821–1934 and a new analysis. *J. Roy. Statist. Soc.*, vol. 140A, p. 411–32.

CAPERON, J.; MEGER, N. 1972. Nitrogen-limited growth of marine phytoplankton. Part II: Uptake kinetics and their role in nutrient limited growth of phytoplankton. *Deep-sea Res. Oceanogr. Abstr.*, vol. 19, p. 619–32.

CARLSON, C. A.; SHEALY, M. H., Jr. 1972. Marking larval largemouth bass with radiostrontium. *J. Fish. Res. Brd Can.*, vol. 29, p. 455–8.

CARTER, J. G. T.; NICHOLAS, W. L. 1978. Uptake of the zinc by the aquatic larvae of *Simulium ornatipes* (Dipera: Nematocera). *Aust. J. Mar. Freshwat. Res.*, vol. 29, p. 299–309.

CASSIE, R. M. 1968. *Sample design*, p. 105–25. Paris, Unesco. (Monogr. oceanogr. Methodol. 2.)

——. 1971. Sampling and statistics. In: W. T. Edmondson and G. G. Winberg (eds.), *A manual on methods for the assessment of secondary production in fresh waters*. Oxford, Blackwell. 358 p. (IBP Handbk 17.)

CASWELL, H. 1976. The validation problem. In: B. C. Patten (ed.), *Systems analysis and simulation in ecology*, vol. IV. New York, N.Y., Academic Press. 593 p.

CHAMBERS, L. A.; TRUDINGER, P. A. 1975. Are thiosulfate and dithionate intermediates in dissimulatory sulfate reduction? *J. Bacteriol.*, vol. 123, p. 36–40.

CHANG, B. D.; PARSONS, T. R. 1975. Metabolic studies on the amphipod *Anisogammarus pugettensis* in relation to its trophic position in the food-web of young salmonids. *J. Fish. Res. Brd Can.*, vol. 32, p. 243–7.

CHANUT, J. P.; POULET, S. A.; COSSA, D.; MARTY, J. C. 1977. Factors analysis of the suspended particulate matter distribution in the St Lawrence Estuarine system. *Can. J. Earth Sci.*, vol. 16, p. 1–23.

CHEN, G. W. 1970. Concepts and utilities of ecological models. *J. Sanit. Eng. Div. Am. Soc. Civ. Eng.*, vol. 96 (SA5), p. 1085–99.

CHMYR, V. D. 1967. Radiocarbon method of determining the production of zooplankton in a natural population. *Dokl. AN Az. SSR*, vol. 173, p. 201–3.

CHROST, R. J.; WAZYK, M. 1978. Primary production and extracellular release by phytoplankton in some lakes in the Masurian Lake District, Poland. *Acta Microbiol. Pol. (A. Microbiol. Gen.)*, vol. 27, p. 63–71.

CLELAND, W. 1967. The statistical analysis of enzyme kinetic data. In: F. F. Nord (ed.), *Advances in enzymology*, 29, p. 1–32. New York, N.Y., Wiley-Interscience.

COLEBROOK, J. M. 1960. Continuous plankton records: methods of analysis 1950–1959. *Bull. Mar. Biol.*, vol. 5, p. 51–64.

——. 1969. Variability in the plankton. *Prog. Oceanogr.*, vol. 5, p. 115–25.

——. 1972. Changes in the distribution and abundance of zooplankton in the North Sea, 1948–1969. *Symp. Zool. Soc. Lond.*, no. 29, p. 203–12.

——. 1975. The continuous plankton recorder survey: computer simulation of some aspects of the design of the survey. *Bull. Mar. Biol.*, vol. 8, p. 143–66.

——. 1977. Annual fluctuations in biomass of taxonomic group of zooplankton in the California Current, 1955–59. *Fish. Bull.*, vol. 75, p. 357–68.

——. 1978. Continuous plankton records: zooplankton and environment, north-east Atlantic and North Sea, 1948–1975. *Oceanol. Acta*, vol. 1, p. 9–23.

COLLETTE, B. B.; EARLE, S. A. 1972. Results of the Tektite program. Ecology of coral reef fishes. *Sci. Bull. Nat. Hist. Mus. Los Ang. Cty*, no. 14. 180 p.

COMITA, G. W. 1968. Oxygen consumption in *Diaptomus*. *Limnol. Oceanogr.*, vol. 13, p. 51–77.

CONOVER, R. J. 1978. Feeding interactions in the pelagic zone. *Rapp. P.-v. Réun. CIEM*, vol. 173, p. 66–76.

CONWAY, E.; HOFF, D.; SMOLLER, J. 1978. Large time behaviour of solutions of systems of nonlinear reaction diffusion equations. *Siam J. Appl. Math.*, vol. 35, p. 1–16.

CRAWFORD, C. C.; HOBBIE, J. W.; WEBB, K. L. 1974. The utilization of dissolved free amino acids by estuarine microorganisms. *Ecology*, vol. 55, p. 551–63.

CRISP, D. 1971. Energy flow measurements. In: N. A. Holme and A. D. McIntyre (eds.), *Methods for the study of marine benthos*, p. 197–279. Oxford, Blackwell Scientific.

CROSSLAND, C. J.; BARNES, D. J. 1974. The role of metabolic nitrogen in coral calcification. *Mar. Biol.*, vol. 28, p. 325–32.

CUMBERLAND, W. G.; ROHDE, C. A. 1977. A multivariate model for growth of populations. *Theor. Popul. Biol.*, vol. 11, p. 127–39.

CUMMINGS, W. C.; BRAHY, B. D.; SPIRES, J. H. 1966. Sound production, schooling and feeding habits of the margate, *Haemulon album* Cuvier, off North Bimini, Bahamas. *Bull. Mar. Sci.*, vol. 16, p. 626–40.

CUSHING, D. H. 1961. On the failure of the Plymouth herring fishery. *J. Mar. Biol. Assoc. U.K.*, vol. 41, p. 799–816.

——. 1971. A comparison of production in temperate seas and the upwelling areas. *Trans. R. Soc. Africa*, vol. 40, no. 1, p. 17–33.

DAVIS, R. W. 1976. Predictability of sea surface temperature and sea level anomalies over the North Pacific. *J. Phys. Oceanogr.*, vol. 6, p. 249–66.

DAYTON, P. K. 1971. Competition, disturbance, and community organization: the provision and subsequent utilization of space in a rocky intertidal community. *Ecol. Monogr.*, vol. 41, p. 351–89.

DEGN, H.; BALSLEV, I.; BOOK, R. (eds.). 1976. *Interdisciplinary symposium on measurement of oxygen*. Amsterdam, Odense University, Elsevier. 276 p.

DEMETRIUS, L. 1977. Macroscopic parameters and ecological systems. *Math. Biosci.*, vol. 36, p. 15–23.

DENBIGH, K. G. 1975. A non-conserved function for organized systems. In: L. Kubat and J. Zeman (eds.), *Entropy and information in science and philosophy*, p. 83–92. Amsterdam, Elsevier.

DENMAN, K. L. 1976. Covariability of chlorophyll and temperature in the sea. *Deep-sea Res. Oceanogr. Abstr.*, vol. 23, no. 6, p. 539–50.

DENMAN, K. L.; MACKAS, D. L. 1978. Collection and analysis of underway data and related physical measurement. In: J. H. Steele (ed.), *Spatial pattern in plankton communities*, p. 85–109. New York, N.Y., Plenum Press.

DENMAN, K. L.; OKUBO, A.; PLATT, T. 1977. The chlorophyll fluctuation spectrum in the sea. *Limnol. Oceanogr.*, vol. 22, p. 1033–8.

DENMAN, K. L.; PLATT, T. 1976. The variance spectrum of phytoplankton in a turbulent ocean. *J. Mar. Res.*, vol. 34, p. 593–601.

DERENBACK, S. B.; LE B. WILLIAMS, P. J. 1974. Autotrophic and bacterial production: fractionation of plankton populations by differential filtration of samples from the English Channel. *Mar. Biol.*, vol. 25, p. 263–9.

DESATY, D.; MCGRATH, R.; VINING, L. C. 1969. Mass spectrometric measurement of ^{15}N enrichment in nitrogen obtained by Dumas combustion. *Anal. Biochem.*, vol. 29, p. 22–30.

DITORO, D. M.; O'CONNOR, D. J.; THONANN, R. V. 1971. In: *Non-equilibrium systems in natural water chemistry*, p. 131–80. American Chemical Society. (Advances in chemistry 106.)

DITORO, D. M.; THOMANN, R. V.; O'CONNOR, D. J.; MANCINI, J. L. 1977. Estuarine phytoplankton biomass models—verification analyses and preliminary applications. In: E. D. Goldberg, I. N. McCave, J. J. O'Brien, J. H. Steele (eds.), *The sea: ideas and observations on progress in the study of the seas*, vol. 6, p. 969–1020. New York, N.Y., John Wiley & Sons.

DUGDALE, R. C. 1967. Nutrient limitation in the sea: dynamics, identification, and significance. *Limnol. Oceanogr.*, vol. 12, p. 685–95.

——. 1975. Biological modelling, I. In: J. C. J. Nihoul (ed.), *Modelling of marine systems*, p. 187–205. Amsterdam, Elsevier.

——. 1977. Nutrient modeling. In: E. D. Goldberg, I. N. McCave, J. J. O'Brien, J. H. Steele (eds.), *The sea: ideas and observations on progress in the study of the seas*, 6, p. 789–806. New York, N.Y., John Wiley & Sons.

DUGDALE, R. C.; GOERING, J. J. 1967. Uptake of new and regenerated forms of nitrogen in primary productivity. *Limnol. Oceanogr.*, vol. 12, p. 196–206.

EATON, J. W.; MOSS, B. 1966. The estimation of numbers and pigment content in epipelic algal populations. *Limnol. Oceanogr.*, vol. 11, p. 584–95.

EDMONDSON, W. T.; WINBERG, G. G. 1971. *A manual on methods for the assessment of secondary production in fresh waters*. Oxford, Blackwell. 358 p. (IBP Handbk 17.)

143

EDWARDS, R. R. C.; FINLAYSON, D. M.; STEELE, J. H. 1972. An experimental study of the oxygen consumption, growth and metabolism of the cod (*Gadus morhua* L.). *J. Exp. Mar. Biol. Ecol.*, vol. 8, p. 299–309.

ERIKSSON, S.; SELLEI, C.; WALLSTRÖM, K. 1977. The structure of the plankton community of the Öregrundsgrepen (Southwest Bosnian Sea). *Helgol. Wiss. Meeresunters*, vol. 30, p. 582–97.

EVANS, G. T.; STEELE, J. H.; KULLENBERG, G. E. B. 1977. A preliminary model of shear diffusion and plankton populations. *Scott. Fish. Res. Rep.*, no. 9 (ISSN 0308 8022).

FALKOWSKI, P. G. 1975. Nitrate uptake in marine phytoplankton: comparison of half-saturation constants from seven species. *Limnol. Oceanogr.*, vol. 20, p. 412–17.

FANKBONER, P. V.; DE BURGH, M. E. 1978. Comparative rates of dissolved organic carbon accumulation by juveniles and pediveligers of the Japanese oyster *Crassostrea gigas* Thunberg. *Aquaculture*, vol. 13, p. 205–12.

FASHAM, M. J. R. 1978a. The statistical and mathematical analysis of plankton patchiness. *Oceanogr. Mar. Biol. Ann. Rev.*, vol. 16, p. 43–79.

——. 1978b. The application of some stochastic processes to the study of plankton patchiness. In: J. H. Steele (ed.), *Spatial pattern in plankton communities*, p. 131–56. New York, N.Y., Plenum Press.

FASHAM, M. J.; PUGH, P. R. 1976. Observations on the horizontal coherence of chlorophyll *a* and temperature. *Deep-sea Res. Oceanogr. Abstr.*, vol. 23, p. 527–38.

FENCHEL, T. 1974. Intrinsic rate of natural increase: the relationship with body size. *Oecologia*, vol. 14, p. 317–26.

FINN, J. T. 1976. Measures of ecosystem structure and function derived from analysis of flows. *J. Theor. Biol.*, vol. 41, p. 535–46.

FISHER, R. A.; CORBETT, A. S.; WILLIAMS, R. B. 1943. The relation between the number of species and the number of individuals in a random sample of an animal population. *J. Anim. Ecol.*, vol. 12, p. 42–58.

FITZGERALD, G. J.; KEENLEYSIDE, M. H. A. 1978. Technique for tagging small fish with I^{131} for evaluation of predation–prey relationships. *J. Fish. Res. Brd Can.*, vol. 35(1), p. 143–54.

FLETT, R. J.; HAMILTON, R. D.; CAMPBELL, N. E. R. 1976. Aquatic acetylene-reduction techniques: solutions to several problems. *Can. J. Microbiol.*, vol. 22, p. 43–51.

FLIERL, G. 1977. The application of linear quasi-geostrophic dynamics to Gulf Stream rings. *J. Phys. Oceanogr.*, vol. 7, p. 365–79.

FRANKENBERG, D.; SMITH, K. L. 1967. Coprophagy in marine animals. *Limnol. Oceanogr.*, vol. 12, p. 443–50.

FRANZISKET, L. 1975. Nitrate uptake by reef corals. *Int. Rev. Gesamt. Hydrobiol.*, vol. 59, p. 1–7.

GARRETT, C.; MUNK, W. 1975. Space-time scales of internal waves: a progress report. *J. Geophys. Res.*, vol. 80, p. 291–7.

GERKING, S. D. (ed.). 1978. *The ecology of freshwater fish production*. New York, N.Y., Halsted Press. 520 p.

GIERE, O. 1975. Population structure, food relations and ecological role of marine oligochaetes, with special reference to meiobenthic species. *Mar. Biol.*, vol. 31, p. 139–56.

GLOVER, R. S. 1967. The continuous plankton recorder survey of the North Atlantic. *Symp. Zool. Soc. Lond.*, vol. 19, p. 189–210.

GOEL, N. S.; MAITRA, S. C.; MONTROLL, E. W. 1971. On the Volterra and other nonlinear models of interacting populations. *Rev. Mod. Phys.*, vol. 43, p. 231–76.

GOEL, N. S.; RICHTER-DYN, N. 1974. *Stochastic models in biology*. New York, N.Y., Academic Press. 269 p.

GOLDBERG, E. D.; MCCAVE, I. N.; O'BRIEN, J. J.; STEELE, J. H. (eds.). 1977. *The sea: ideas and observations on progress in the study of the seas*, 6. New York, N.Y., John Wiley & Sons. 1048 p.

GOLLEY, F. B. 1974. Structural and functional properties as they influence ecosystem stability. *Proc. First Int. Congr. Ecol.*, p. 97–102. The Hague, Centre for Agricultural Publishing and Documentation, Wageningen.

GRANT, H. L.; STEWART, R. W.; MOILLET, A. 1962. Turbulence spectra from a tidal channel. *J. Fluid Mech.*, vol. 12, p. 241–3.

GRASSHOFF, K.; JOHANNSEN, H. 1972. A new sensitive and direct method for the automatic determination of ammonia in seawater. *J. Cons. CIEM*, vol. 34, p. 516–21.

HAEDRICH, A. L.; ROWE, G. T. 1977. Megafaunal biomass in the deep sea. *Nature* (Lond.), vol. 269, p. 141–2.

HALFON, E. A. 1979. *Theoretical systems ecology*. New York, N.Y., Academic Press. 516 p.

144

HALL, A. S.; DAY, J. W., Jr. 1977. *Ecosystem modelling in theory and practice: an introduction with case histories*. New York, N.Y., J. Wiley & Sons. 684 p.

HAMNER, W. M.; MADIN, L. P.; ALLDREDGE, A. L; GILMER, R. W.; HAMNER, P. P. 1975. Underwater observations of gelatinous zooplankton sampling problems, feeding, biology and behaviour. *Limnol. Oceanogr.*, vol. 20, p. 907–17.

HANNON, B. 1973. The structure of ecosystems. *J. Theor. Biol.*, vol. 41, p. 535–46.

HARGRAVE, B. T.; PHILLIPS, G. A. 1976. D.D.T. residues in benthic invertebrates and demersal fish in St Margaret's Bay, Nova Scotia. *J. Fish. Res. Brd Can.*, vol. 33, p. 1692–8.

HARRIS, E. 1957. Radiophosphorus metabolism in zooplankton and micro-organisms. *Can. J. Zool.*, vol. 35, p. 769–82.

HARVEY, H. W. 1950. On the production of living matter in the sea off Plymouth. *J. Mar. Biol. Assoc. U.K.*, vol. 29, p. 97–137.

HARVEY, W. A.; CAPERON, J. 1976. The rate of utilization of urea, ammonium, and nitrate by natural populations of marine phytoplankton in an eutrophic environment. *Pac. Sci.*, vol. 30, p. 329–40.

HASTINGS, J. W. 1955. The effect of oxygen concentration upon the luminescence of bacterial extracts. *Anat. Rec.*, vol. 122, p. 458.

HASTINGS, J. W.; GIBSON, Q. H. 1963. Intermediates in the bioluminescent oxidation of reduced flavin mononucleotide. *J. Biol. Chem.*, vol. 238, p. 2537–54.

HAURY, L. R.; McGOWAN, J. A.; WIEBE, P. H. 1978. Patterns and processes in the time-space scales of plankton distributions. In: J. H. Steele (ed.), *Spatial pattern in plankton communities*, p. 277–328. New York, N.Y., Plenum Press.

HEARON, J. Z. 1963. Theorems on linear systems. *Ann. N.Y. Acad. Sci.*, vol. 108, p. 36–68.

HEDGPETH, J. M. 1977. Models and muddles. Some philosophical observations. *Helgol. Wiss. Meeresunters.*, vol. 30, p. 92–104.

HIGUCHI, M.; MUKADE, A. 1976. Distribution of the bound form tritium in the whole body of brine ships, *Artemia salina*, reared in the seawater containing tritiated water. *Radioisotopes*, vol. 25, p. 51–6.

HILL, B. J. 1976. Natural food, foregut clearance-rate and activity of the crab *Scylla serrata*. *Mar. Biol.*, vol. 34, p. 109–16.

HIRATA, H.; FUKAO, T. 1977. A model of mass and energy flow in ecosystems. *Math. Biosci.*, vol. 33, p. 321–34.

HOLME, N. A.; McINTYRE, A. D. 1971. *Methods for the study of marine benthos*. Oxford, Blackwell. 334 p. (IBP Handbk 16.)

HOMER, M.; KEMP, W. M. 1977. Functional analysis of complex estuarine food-webs. (Unpubl. MS.)

HORWOOD, J. W. 1978. Observations on spatial heterogeneity of surface chlorophyll in one and two dimensions. *J. Mar. Biol. Assoc. U.K.*, vol. 58, p. 487–502.

HULTBORN, R. 1972. A sensitive method for measuring oxygen consumption. *Anal. Biochem.*, vol. 47, p. 442–50.

HUTCHINSON, G. E. 1959. Homage to Santa Rosalia, or Why are there so many kinds of animals? *Am. Nat.*, vol. 93, p. 145–59.

IKEDA, T. 1974. Nutritional ecology of marine zooplankton. *Mem. Fac. Fish. Hokkaido Univ.*, vol. 22, p. 1–97.

ISAACS, J. D. 1972. Unstructured marine food-webs and 'pollutant analogues'. *Fish. Bull.*, vol. 70, p. 1053–9.

——. 1973. Potential trophic biomasses and trace substance concentration in unstructured marine food-webs. *Mar. Biol.*, vol. 22, p. 97–104.

IVERSON, R. L.; CURL, H. C.; SAUGEN, J. L. 1974. Simulation model for wind-driven summer phytoplankton dynamics in Auke Bay, Alaska. *Mar. Biol.*, vol. 28, p. 169–77.

IVLEV, V. S. 1945. The biological productivity of waters. *Usp. Sovrem. Biol.*, vol. 19, p. 98–120. *Fish. Res. Brd Can. Trans.*, ser. 394, from *J. Fish. Res. Brd Can.*, vol. 23, p. 1727–44.

JAMART, B. M.; WINTER, D. F.; BANSE, K.; ANDERSON, G. C.; LAM, R. K. 1978. A theoretical study of phytoplankton growth and nutrient distribution in the Pacific Ocean off the North Western U.S. Coast. *Deep-sea Res. Oceanogr. Abstr.*, vol. 24, p. 753–73.

JANNSON, B. O. 1972. Ecosystem approach to the Baltic problem. *Bull. Ecol. Res. Comm.*, vol. 16, p. 1–82. (Swed. Nat. Sci. Res. Coun.)

JASSBY, A. D.; PLATT, T. 1976. Mathematical formulation of the relationship between photosynthesis and light for phytoplankton. *Limnol. Oceanogr.*, vol. 21, no. 4, p. 540–7.

JENKINS, G. M.; WATTS, D. G. 1968. *Spectral analysis and its applications*. San Francisco, Calif., Holden-Day. 523 p.

145

JENNINGS, J. B.; GIBSON, R. 1969. Observations on the nutrition of seven species of Rhynchocoelan worms. *Biol. Bull.*, vol. 136, p. 405–33.

JERLOV, N. G. 1961. Optical measurements in the eastern North Atlantic. *Göteborgs Kgl. Vetenskaps-Vitterhets-Samhall. Handl.*, ser. B., 8, p. 1–39.

JITTS, H. R.; MOREL, A.; SAIJO, Y. 1976. The relation of oceanic primary production to available photosynthetic irradiance. *Aust. J. Mar. Freshwat. Res.*, vol. 27, p. 441–54.

JOHANNES, R. E. 1964. Uptake and release of phosphorus by a benthic marine amphipod. *Limnol. Oceanogr.*, vol. 9, p. 235–42.

JOHNSON, R. G. 1970. Variations in diversity within benthic marine communities. *Am. Nat.*, vol. 104, p. 285–300.

JONES, N. S.; KAIN, J. M. 1967. Subtidal algal colonization following removal of *Echinus*. *Helgol. Wiss. Meeresunters.*, vol. 15, p. 460–6.

JORGENSEN, S. E.; MEJER, H.; FRILIS, M. 1978. Examination of a lake model. *Ecol. Modelling*, vol. 4, p. 253–78.

KAHN, L. F.; HAYNES, H. H. 1970. Undersea concrete habitats. *Ocean Ind.*, vol. 5, p. 27–9.

KELLY, J. C. 1976. Sampling the sea. In: H. Cushing and J. J. Walsh (eds.), *The ecology of the seas*, p. 361–87. Philadelphia, Pa, W. B. Saunders.

KERNER, E. H. 1957. A statistical mechanics of interacting biological species. *Bull. Math. Biophys.*, vol. 19, p. 121–46.

——. 1959. Further considerations on the statistical mechanics of biological associations. *Bull. Math. Biophys.*, vol. 21, p. 217–55.

——. 1961. On the Lotka-Volterra principle. *Bull. Math. Biophys.*, vol. 23, p. 141–57.

——. 1962. Gibbs ensemble and biological ensemble. *Ann. N.Y. Acad. Sci.*, vol. 96, p. 975–84.

——. 1964. Dynamical aspects of kinetics. *Bull. Math. Biophys.*, vol. 26, p. 333–49.

——. 1972. *Gibbs ensemble: biological ensemble*. New York, N.Y., Gordon & Breach. 167 p.

KERR, S. R. 1974a. Theory of size distribution in ecological communities. *J. Fish. Res. Brd Can.*, vol. 31, p. 1859–62.

——. 1974b. Structural analysis of aquatic communities. *Proc. First Int. Congr. Ecol.*, p. 69–74. The Hague, Centre for Agricultural Publishing and Documentation, Wageningen.

KIERSTEAD, H.; SLOBODKIN, L. B. 1953. The size of water masses containing plankton blooms. *J. Mar. Res.*, vol. 12, no. 1, p. 141–7.

KINSEY, D. W. 1972. Preliminary observations on community metabolism and primary productivity of the pseudo-atoll reef at One Tree Island, Great Barrier Reef. In: C. Kukandan and C. S. G. Pillai (eds.), *Proceedings of the symposium on corals and coral reefs, 1969*, p. 13–32. Cochin, Mar. Biol. Assoc. of India.

KITCHEN, J. C.; MENZIES, D.; PAK, H.; ZANEVELD, R. V. 1975. Particle size distributions in a region of coastal upwelling analyzed by characteristic vectors. *Limnol. Oceanogr.*, vol. 20, p. 775–83.

KITTEL, C. 1958. *Elementary statistical physics*. New York, N.Y., J. Wiley & Sons. 228 p.

KLEINEN HAMMANS, J. W.; HENDRIKS, G. M.; TEERLINK, T. 1977. Light dependent oxygen uptake by the blue green alga *Anacystis nidulans*. *Biochem. Biophys. Res. Commun.*, vol. 74, p. 1560–5.

KOLMOGOROV, A. N. 1941. The local structure of turbulence in an incompressible viscous fluid for very large Reynolds number. *Dokl. AN Az. SSR*, vol. 30, p. 299–303.

KREDICH, N. M.; TOMKINS, G. M. 1966. The enzymic synthesis of L-cysteine in *Escherischia coli* and *Salmonella typhimurium*. *J. Biol. Chem.*, vol. 241, p. 4955–65.

KREMER, J. N.; NIXON, S. W. 1978. *A coastal marine ecosystem—simulation and analysis*. New York, N.Y., Springer-Verlag. 217 p.

KULLENBERG, G. 1974. An experimental and theoretical investigation of the turbulent diffusion in the upper layer of the sea. *Univ. Copenhagen Rep.*, no. 25. 212 p.

LAI, D. Y.; RICHARDSON, P. L. 1977. Distribution and movement of Gulf Stream rings. *J. Phys. Oceanogr.*, vol. 7, p. 670–83.

LANE, P. A. 1978. Role of invertebrate predation in structuring zooplankton communities. *Verh. Int. Ver. Theor. Agnew. Limnol.*, vol. 20, p. 480–5.

LANE, P. A.; KLUG, M. J.; LOUDEN, L. 1976. Measuring invertebrate predation *in situ* on zooplankton assemblages. *Trans. Am. Microscop. Soc.*, vol. 95, p. 143–55.

LANE, P.; LEVINS, R. 1977. The dynamics of aquatic systems. II. The effects of nutrient enrichment on model plankton communities. *Limnol. Oceanogr.*, vol. 22, no. 3, p. 454–71.

LANGE, G. D.; HURLEY, A. C. 1975. A theoretical treatment of unstructured food-webs. *Fish. Bull.*, vol. 73, p. 378–81.

146

LAST, J. M. 1978. The food of four species of Pleuronectiform larvae in the Eastern English Channel and Southern North Sea. *Mar. Biol.*, vol. 45, p. 359–68.

LEGAND, M.; BOURRET, P.; FOURMANIOR, P.; GRANDPERIN, R.; GUEREDRAT, J. A.; MICHEL, A.; RANCUREL, P.; REPELIN, R.; ROGER, C. 1971. On trophic relationships at higher levels of the food chain in the tropical Pacific Ocean. In: *Proc. 12th Pacif. Sci. Congr.*, 1, p. 159. Canberra, Australian Academy of Science for the Pacific Science Association.

LEIGH, E. G., Jr. 1965. On the relation between the productivity, biomass, diversity and stability of a community. *Proc. NAS*, vol. 53, p. 777–83.

——. 1968. The ecological role of Volterra's equations. In: M. Gerstenhaber (ed.), *Some mathematical problems in ecology*, p. 1–61. Providence, R.I., American Mathematical Society.

LEKAN, J. F.; WILSON, R. E. 1978. Spatial variability of phytoplankton biomass in the surface waters of Long Island. *Estuar. Coast. Mar. Sci.*, vol. 6, p. 239–51.

LEONTIEF, W. 1951. *The structure of the American economy, 1919–1939.* 2nd ed. New York, N.Y., Oxford University Press. 264 p.

LEVINE, S. H. 1977. Exploitation interactions and the structure of ecosystems. *J. Theor. Biol.*, vol. 69, p. 345–55.

——. 1980. Some measures of trophic structure. Submitted to *J. Theor. Biol.*

LEVINS, R. 1974. The qualitative analysis of partially specialized systems. *Ann. N.Y. Acad. Sci.*, vol. 231, p. 123–38.

LIMER 1975 EXPEDITION TEAM. 1976. Metabolic processes of coral reef communities at Lizard Island, Queensland. *Search* (Syd.), vol. 7, p. 463–8.

LINDEMAN, R. L. 1942. The trophic-dynamic aspect of ecology. *Ecology*, vol. 23, p. 399–418.

LITTLER, M. M. 1973. The productivity of Hawaiian Fringing Reef crustose *Coral-linaceae* and an experimental evaluation of production methodology. *Limnol. Oceanogr.*, vol. 18, p. 946–52.

LITTLER, M. M.; MURRAY, S. N. 1974. The primary productivity of marine macrophytes from a rocky intertidal community. *Mar. Biol.*, vol. 27, p. 131–5.

LONGHURST, A. R. 1978. Ecological models in estuarine management. *Ocean Mgmt*, vol. 4, p. 287–302.

LONGMUIR, I. L.; KNOPP, J. A. 1973. In: H. I. Bicher and D. F. Bruley (eds.), *Advances in experimental medicine and biology*, 37A, p. 55–7. New York, N.Y., Plenum Press.

LORENZ, E. N. 1963. Deterministic nonperiodic flow. *J. Atmos. Sci.*, vol. 20, p. 130–41.

——. 1969. The predictability of a flow which possesses many scales of motion. *Tellus*, vol. 21, no. 3, p. 291–307.

LOTKA, A. J. 1925. *Elements of physical biology.* Baltimore, Md, Williams & Wilkins. 460 p.

MCALLISTER, C. D.; PARSONS, T. R.; STEPHENS, K.; STRICKLAND, J. D. H. 1961. Measurements of primary production in coastal sea water using a large volume plastic sphere. *Limnol. Oceanogr.*, vol. 6, p. 237–58.

MCCARTHY, J. J.; TAYLOR, W. R.; TAFT, J. L. 1977. Nitrogenous nutrition of the plankton in the Chesapeake Bay. 1. Nutrient availability and phytoplankton preferences. *Limnol. Oceanogr.*, vol. 22, p. 996–1011.

MACKAS, D. L. 1977. *Horizontal spatial variability and covariability of marine phytoplankton and zooplankton.* Halifax (Canada), Dalhousie University. (Ph.D. thesis.)

MACKAY, D. W. 1973. A model for levels of dissolved oxygen in the Clyde. *Wat. Pollut. Control Res. Tech. Pap.*, vol. 13, p. 85–94.

MCKINLEY, K.; WETZEL, R. G. 1977. Tritium oxide uptake by algae: an independent measure of phytoplankton photosynthesis. *Limnol. Oceanogr.*, vol. 22, p. 377–80.

MCMAHON, J.; RIGLER, F. H. 1963. Mechanisms regulating the feeding rate of *Daphnia magna* Straus. *Can. J. Zool.*, vol. 41, p. 321–32.

MCMURTRIE, R. 1978. Persistence and stability of single-species and prey–predator systems in spatially heterogeneous environments. *Math. Biosci.*, vol. 39, p. 11–51.

MANN, K. H. 1969. Dynamics of aquatic ecosystems. In: J. B. Cragg (ed.), *Ad. Ecol. Res.*, vol. 6, p. 1–81. New York, N.Y., Academic Press.

——. 1972. Ecological energetics of the sea-weed zone in a marine bay on the Atlantic coast of Canada, II. Productivity of the sea-weeds. *Mar. Biol.*, vol. 14, p. 199–209.

——. 1974. Comparison of freshwater and marine systems: the direct and indirect effects of solar energy on primary and secondary production. In: *Proc. First Int. Congr. Ecol.*, p. 168–73. The Hague, PUPOC (Wageningen).

——. 1975. Relationship between morphometry and biological functioning in three coastal inlets of Nova Scotia. In: L. E. Cronin (ed.), *Estuarine Research*, vol. 1, p. 634–44. New York, N.Y., Academic Press.

147

——. 1977. Destruction of kelp-beds by sea urchins: a cyclical phenomenon or irreversible degradation? *Helgol. Wiss. Meeresunters.*, vol. 30, p. 455–67.

MARGALEF, R. 1968. *Perspectives in ecological theory.* Chicago, Ill., Chicago Press. 111 p.

——. 1978. Life-forms of phytoplankton as survival alternatives in an unstable environment. *Oceanol. Acta*, vol. 1, p. 493–509.

MARSHALL, N.; OVIATT, C. A.; SKAUEN, D. M. 1971. Productivity of the benthic microflora of shoal estuarine environments in Southern New England. *Int. Rev. Gesamt. Hydrobiol.*, vol. 56, p. 947–56.

MASON, C. F. 1977. Populations and production of benthic animals in two contrasting shallow lakes in Norfolk. *J. Anim. Ecol.*, vol. 46, p. 147–72.

MAY, R. M. 1973. *Stability and complexity in model ecosystems.* Princeton, N.J., Princeton University Press. 235 p.

MAY, R. M. (ed.). 1976. *Theoretical ecology: principles and applications.* Oxford, Blackwell Scientific. 317 p.

MAY, R. M.; OSTER, G. F. 1976. Bifurcations and dynamic complexity in simple ecological models. *Am. Nat.*, vol. 110, p. 573–99.

MAYNARD SMITH, J. 1974. *Models in ecology.* Cambridge, Cambridge University Press. 146 p.

MAYZAUD, P.; POULET, S. A. 1978. The importance of the time factor in the response of zooplankton to varying concentrations of naturally occurring particulate matter. *Limnol. Oceanogr.*, vol. 23, p. 1144–54.

MELLOR, G. L.; DURBIN, P. A. 1975. The structure and dynamics of the ocean surface mixed layer. *J. Phys. Oceanogr.*, vol. 5, p. 718–28.

MENZEL, D. W.; CORWIN, N. 1965. The measurement of total phosphorus in seawater based on the liberation of organically bound fractions by persulfate oxidation. *Limnol. Oceanogr.*, vol. 10, p. 280–2.

MODE GROUP, THE. 1978. The Mid-Ocean dynamics experiment. *Deep-Sea Res. Oceanogr. Abstr.*, vol. 25, p. 859–910.

MOORE, S.; STEIN, W. H. 1954. A modified ninhydrin method for the photometric determinations of amino acids and related compounds. *J. Biol. Chem.*, vol. 211, p. 907–13.

MORTIMER, C. H. 1975. Modelling of lakes as physico-biochemical systems-present limitations and needs. In: J. C. J. Nihoul (ed.), *Modelling of marine systems*, p. 217–32. Amsterdam, Elsevier.

MULHOLLAND, R. J. 1975. Stability analysis of the response of ecosystems to perturbations. In: S. A. Levin (ed.), *Ecosystems: analysis and prediction*, p. 166–81. Alta, Utah. (SIAMS-SIMS Res. Applic. Conf. Ecosyst.)

MULHOLLAND, R. J.; SIMMS, G. S. 1976. Control theory and the regulation of ecosystems. In: B. C. Patten (ed.), *Systems analysis and simulation in ecology*, p. 373–89. New York, N.Y., Academic Press.

MÜLLER, D.; KNOPP, H. 1971. Zur messung der primarproduktion an der biogenen beluftung in Fliessgewassern. *Int. Rev. Gesamt. Hydrobiol.*, vol. 56, p. 49–67.

MULLIN, M. M. 1965. Size fractionation of particulate organic carbon in the surface waters of the western Indian Ocean. *Limnol. Oceanogr.*, vol. 10, p. 459–62.

——. 1966. Selective feeding by calanoid copepods from the Indian Ocean. In: H. Barnes (ed.), *Some contemporary studies in marine science*, p. 545–54. London, Allen & Unwin.

MULLIN, M. M.; BROOKS, E. R. 1967. Laboratory culture, growth rate, and feeding behaviour of a planktonic marine copepod. *Limnol. Oceanogr.*, vol. 12, p. 657–66.

MUNK, W. H.; ANDERSON, E. R. 1948. Notes on a theory of the thermocline. *J. Mar. Res.*, vol. 7, p. 276–95.

MURPHY, G. I. 1966. Population biology of the Pacific Sardine (*Sardinops caerulea*). *Proc. Calif. Acad. Sci.*, vol. 34, p. 1–84.

MURPHY, J.; RILEY, J. P. 1962. A modified single solution method for the determination of phosphate in natural waters. *Anal. Chim. Acta*, vol. 27, p. 31–6.

MYRBERG, A. A., Jr. 1972. Social dominance and territoriality in the bicolour damselfish, *Eupomacentrus partitus* (Poey) (Pisces: Pomacentridae). *Behaviour*, vol. 41, p. 207–31.

——. 1973a. Underwater television—a tool for the marine biologist. *Bull. Mar. Sci.*, vol. 23, p. 824–36.

——. 1973b. Ethology of the bicolour damsel fish, *Eupomacentrus partitus* (Pisces: Pomacentridae). A comparative analysis of laboratory and field behaviour. *Anim. Behav. Monogr.*, vol. 5, p. 199–283.

MYRBERG, A. A., Jr.; THRESHER, R. E. 1974. Interspecific aggression and its relevance to the concept of territoriality in reef fishes. *Am. Zool.*, vol. 14, p. 79–94.

NELSON, D. M.; GOERING, J. J. 1977. A stable isotope tracer method to measure silicic acid uptake by marine phytoplankton. *Anal. Biochem.*, vol. 78, p. 139–47.

NEWELL, B. S.; MORGAN, B.; CUNDY, J. 1967. The determination of urea in seawater. *J. Mar. Res.*, vol. 25, p. 201–2.

NICOLIS, G.; PRIGOGINE, I. 1977. *Self organization in non-equilibrium systems.* New York, N.Y., Wiley-Interscience. 491 p.

NICOTRI, M. E. 1977. Grazing effects of four marine intertidal herbivores on the microflora. *Ecology*, vol. 58, p. 1020–32.

NIHOUL, J. C. J. (ed.). 1975. *Modelling of marine systems.* Amsterdam, Elsevier Scientific. 272 p. (Elsevier Oceanogr. ser. 10.)

NIHOUL, J. C. J. 1976. A small interdisciplinary mathematical model applied to the southern bight of the North Sea. *Ecol. Modelling*, vol. 2, p. 3–17.

NIILER, P. P.; KRAUS, E. B. 1977. One dimensional models of the upper ocean. In: E. B. Kraus (ed.), *Modelling and prediction of the upper layers of the ocean*, p. 143–72. Oxford, Pergamon Press.

NIMMO, D. R.; WILSON, A. J., Jr.; BLACKMAN, R. R. 1970. Localization of DDT in the body organs of pink and white shrimp. *Bull. Environ. Contam. Toxicol.*, vol. 5, p. 333–40.

NIVAL, P.; NIVAL, S.; PALAZZOLI, I. 1972. Données sur la respiration de différents organismes communs dans le plancton de Villefranche-sur-Mer. *Mar. Biol.*, vol. 17, p. 63–76.

NORTH, B. B. 1975. Primary amines in California coastal waters; utilisation by phytoplankton. *Limnol. Oceanogr.*, vol. 20, p. 20–7.

O'BRIEN, J. J.; WROBLEWSKI, J. S. 1973. A simulation of the mesoscale distribution of the lower marine trophic levels off West Florida. *Invest. Pesq.*, vol. 37, p. 193–244.

ODUM, E. P. 1953. *Fundamentals of ecology.* Philadelphia, Pa, Saunders. 384 p.

——. 1959. *Fundamentals of ecology.* 2nd ed. Philadelphia, Pa, Saunders. 546 p.

——. 1969. The strategy of ecosystem development. *Science*, vol. 164, p. 262–70.

ODUM, H. T. 1957. Trophic structure and productivity of Silver Springs, Florida. *Ecol. Monogr.*, vol. 27, p. 55–112.

——. 1971. *Environment, power and society.* New York, N.Y., John Wiley & Sons. 331 p.

——. 1972. An energy circuit language for ecological and social systems: its physical basis. In: B. C. Patten (ed.), *Systems analysis and simulations in ecology*, II. New York, N.Y., Academic Press. 592 p.

ODUM, H. T.; BURKHOLDER, P.; REVERS, P. 1959. Measurements of productivity of turtle grass flats, reefs and the Bahia Fosforescente of Southern Puerto Rico. *Publ. Inst. Mar. Sci. Univ. Texas*, vol. 6, p. 159–70.

ODUM, H. T.; HOSKIN, C. M. 1958. Comparative studies in the metabolism of marine waters. *Publ. Inst. Mar. Sci. Univ. Texas*, vol. 6, p. 159–70.

ODUM, H. T.; PINKERTON, R. C. 1955. Time's speed regulator: the optimum efficiency for maximum power output in physical and biological systems. *Am. Sci.*, vol. 43, p. 331–43.

OKUBO, A. 1971. Ocean diffusion diagrams. *Deep-sea Res. Oceanogr. Abstr.*, vol. 18, p. 789–802.

——. 1974. Diffusion-induced instability in model ecosystems: another possible explanation of patchiness. *Tech. Rep.* Baltimore, Md, Chesapeake Bay Institute, Johns Hopkins University. 17 p. (No. 86.)

——. 1978. Horizontal dispersion and critical scales for phytoplankton patches. In: J. H. Steele (ed.), *Spatial pattern in plankton communities*, 3, p. 21–36. New York, N.Y., Plenum Press. (Nato Conf. ser.: IV. Mar. Sci.)

OLSEN, S. 1967. Determination of orthophosphate in water. In: H. L. Golterman and R. S. Clymo (eds.), *Chemical environment in the aquatic habitat*, p. 63–105. Amsterdam, North Holland.

PAFFENHÖFFER, G. A.; STRICKLAND, J. D. H. 1970. A note on the feeding of *Calanus helgolandicus* on detritus. *Mar. Biol.*, vol. 5, p. 97–9.

PALOHEIMO, J. E.; DICKIE, L. M. 1965. Food and growth of fishes. I. A growth curve derived from experimental data. *J. Fish. Res. Brd Can.*, vol. 22, p. 521–42.

——; ——. 1966. Food and growth of fishes. III. Relations among food, body size, and growth efficiency. *J. Fish. Res. Brd Can.*, vol. 23, no. 8, p. 1209–48.

PANKHURST, R. C. (ed.). 1964. *Dimensional analysis and scale factors.* London, Chapman & Hall. 151 p.

PARK, R. A.; O'NEILL, R.; SHUGART, H.; GOLDSTEIN, R.; BOOTH, R.; MANKIN, J. B.; KOONCE, J.; BLOOMFIELD, J. 1974. A generalized model for simulating lake ecosystems. *Simulation*, vol. 23, no. 2, p. 33–50.

PARSONS, T. R. 1969. The use of particle size spectra in determining the structure of a plankton community. *J. Oceanogr. Soc. Jap.*, vol. 25, p. 172–81.

PARSONS, T. R.; LEBRASSEUR, R. J.; FULTON, J. D. 1967. Some observations on the dependence of zooplankton grazing on the cell size and concentration of phytoplankton blooms. *J. Oceanogr. Soc. Jap.*, vol. 23, p. 10–17.

PARSONS, T. R.; SEKI, H. 1969. A short review of some automated techniques for the detection and characterization of particles in sea water. *Bull. Jap. Soc. Fish. Oceanogr.*, November, p. 173–7. (Spec. no.)

PARSONS, T. R.; STRICKLAND, D. H. 1962. On the production of particulate organic carbon by heterotrophic processes in sea water. *Deep-sea Res. Oceanogr. Abstr.*, vol. 8, p. 211–22.

PATTEN, B. C. 1971. *Systems analysis and simulation in ecology*, vol. I. New York, N.Y., Academic Press. 607 p.

——. 1972. *Systems analysis and simulation in ecology*, vol. II. New York, N.Y., Academic Press. 592 p.

——. 1975. *Systems analysis and simulation in ecology*, vol. III. New York, N.Y., Academic Press. 601 p.

PATTEN, B. C.; EGLOFF, D. A.; RICHARDSON, T. H. 1975. Total ecosystem model for a cove in Lake Texoma. In: B. C. Patten (ed.), *Systems analysis and simulation ecology*, vol. III, p. 206–423. New York, N.Y., Academic Press.

PENTREATH, R. J. 1977. Radionuclides in marine fish. *Oceanogr. Mar. Biol. Ann. Rev.*, vol. 15, p. 365–460.

PERONI, C. 1970. The possible role of marine bacteria in the recycling of radionuclides in seawater. *Rev. Int. Océanogr. Méd.*, vol. 20, p. 53–77.

PETIPA, T. S.; PAVLOVA, E. V.; MIRONOV, G. N. 1970. The food web structure, utilization and transport of energy by trophic levels in the planktonic communities. In: J. H. Steele (ed.), *Marine food chains*, p. 142–67. Edinburg, Oliver & Boyd. 522 p.

PETROCELLI, S. F.; ANDERSON, J. W.; HANKS, A. R. 1975a. Biomagnification of dieldrin residues by food-chain transfer from clams to blue-crabs under controlled conditions. *Bull. Environ. Contam. Toxicol.*, vol. 13, p. 108–16.

PETROCELLI, S. F.; ANDERSON, J. W.; HANKS, A. R. 1975b. Controlled food-chain transfer of dieldrin residues from phytoplankters to clams. *Mar. Biol.*, vol. 31, p. 215–18.

PICCARD, J. 1966. The future of deep ocean exploration. *Oceanol. Int.*, vol. 1, p. 51–3.

PILSON, M. E. Q.; BETZER, S. B. 1973. Phosphorus flux across a coral reef. *Ecology*, vol. 54, p. 581–8.

PINGREE, R. D. 1978. Mixing and stabilization of phytoplankton distributions on the Northwest European continental shelf. In: J. H. Steele (ed.), *Spatial pattern in plankton communities*, p. 181–220. New York, N.Y., Plenum Press.

PLATT, T. 1972. Local phytoplankton abundance and turbulence. *Deep-sea Res. Oceanogr. Abstr.*, vol. 19, p. 183–7.

PLATT, T.; DENMAN, K. L. 1975. Spectral analysis in ecology. *Annu. Rev. Ecol. Syst.*, vol. 6, p. 189–210.

PLATT, T.; DENMAN, K. 1977. Organization in the pelagic ecosystem. *Helgol. Wiss. Meeresunters.*, vol. 30, p. 575–81.

——;——. 1978. The structure of pelagic marine ecosystems. *Rapp. P.-v. Réun. CIEM*, vol. 173, p. 60–5.

PLATT, T.; DENMAN, K. L.; JASSBY, A. D. 1977. Modeling the productivity of phytoplankton. In: E. D. Goldberg, I. N. McCave, J. J. O'Brien and J. H. Steele (eds.), *The sea: ideas and observations of progress in the study of the seas*, vol. 6, p. 807–56. New York, N.Y., John Wiley & Sons.

PLATT, T.; GALLEGOS, C.; HARRISON, W. G. 1980. Photoinhibition of photosynthesis in natural assemblages of marine phytoplankton. *J. Mar. Res.*, vol. 38, p. 687–701.

POLLARD, R. T. 1977. Observations and models of the structure of the upper ocean. In: E. B. Kraus (ed.), *Modelling and prediction of the upper layers of the ocean*, p. 102–17. Oxford, Pergamon Press.

POMEROY, L. R. 1974. The ocean's food-wed, a changing paradigm. *BioScience*, vol. 24, p. 499–504.

POMEROY, L. R.; PILSON, M. E. Q.; WIEBE, W. J. 1974. Tracer studies of the exchange of phosphorus between reef water and organisms on the windward reef of Eniwetok atoll. *Proc. 2nd Int. Coral Reef Symp.*, p. 87–96. Brisbane, Great Barrier Reef Committee.

POOLE, R. W. 1976. Empirical multivariate autogressive equation predictors of the fluctuations of interacting species. *Math. Biosci.*, vol. 28, p. 81–97.

PORTER, J. W. 1974. Zooplankton feeding by the Caribbean reef-building coral *Montastrea caverno-sa*. In: A. M. Cameron, B. M. Campbell, A. B. Cribb, R. Endean, J. S. Jell, D. A. Jones, P. Mathei, F. H. Talbot (eds.), *Proc. 2nd Int. Coral Reef Symp.*, 1, p. 111–26. Brisbane, Great Barrier Reef Committee. 630 p.

PORTER, K. G. 1976. Enhancement of algal growth and productivity by grazing zooplankton. *Science*, vol. 192, p. 1332–4.

POULET, S. A. 1976. Feeding of *Pseudocalanus minutus on* living and non-living particles. *Mar. Biol.*, vol. 34, p. 117–25.

POULET, S. A.; MARSOT, P. 1978. Chemosensory grazing by marine calanoid copepods (Anthropoda: Crustacea). *Science*, vol. 200, p. 1403–5.

PRITZLAFF, J. A. 1970. Submersibles, submersibles, submersibles. *Oceanol. Int.*, vol. 5, p. 38–43.

PUGH, P. R. 1978. The application of particle counting to an understanding of the small-scale distribution of plankton. In: J. H. Steele (ed.), *Spatial pattern in plankton communities*, 3, p. 111–29. New York and London, Plenum Press. (Nato Conf. Ser.: IV. Mar. Sci.)

RADACH, G. 1979. (In press.) Preliminary simulations of the phytoplankton and phosphate dynamics during FLEX '76 with a simple 2-component-model. *Meteor. Forschungsergeb.* (*A. Allg. Phys. Chem. Meeres*).

RADACH, G.; MAIER-REIMER, E. 1975. The vertical structure of phytoplankton growth dynamics. A mathematical model. *Mem. Soc. Roy. Sci. Liège*, vol. 7, p. 113–46.

RADACH, G.; MAIER-REIMER, E.; BROCKMANN, U. 1979. Mathematical models as tools for investigating the flow of matter through the lower trophic levels of the marine ecosystem. *I.C.E.S.C.M.*, L. 12. Warsaw.

RADFORD, P. J.; Joint, I. R. 1980. The application of ecosystem model to the Bristol Channel and Estuary. *Inst. Wat. Pollut. Control Ann. Conf.* (Conference paper 7.) (*Water pollution research*, in press.)

RESCIGNO, A; SEGRE, G. 1966. *Drug and tracer kinetics*. Waltham, Mass., Blarsdell. 1968 p.

RICE, C. P.; SIKKA, H. 1973a. Fate of dieldrin in selected species of marine algae. *Bull. Environ. Contam. Toxicol.*, vol. 9, p. 116–23.

——;——. 1973b. Uptake and metabolism of DDT by six species of marine algae. *J. Agric. Food Chem.*, vol. 21 (2), p. 148–52.

RICHERSON, P. J.; POWELL, T. M.; LEIGH-ABBOTT, M. R.; COIL, J. A. 1978. Spatial heterogeneity in closed basins. In: J. H. Steele (ed.), *Spatial pattern in plankton communities*, p. 239–76. New York, N.Y., Plenum Press.

RICHMAN, S.; ROGERS, J. N. 1969. The feeding of *Calanus helgolandicus* on synchronously growing populations of the marine diatom *Ditylum brightwellii*. *Limnol. Oceanogr.*, vol. 14, p. 701–9.

RIEPER, M. 1978. Bacteria as food for marine harpacticoid copepods. *Mar. Biol.*, vol. 45, p. 337–45.

RIGLER, F. H. 1961. The uptake and release of inorganic phosphorus by *Daphina magna* Straus. *Limnol. Oceanogr.*, vol. 6, p. 165–74.

RILEY, G. A. 1963. Organic aggregates in seawater and the dynamics of their formation and utilization. *Limnol. Oceanogr.*, vol. 8, p. 372–81.

RILEY, G. A.; STOMMEL, H.; BUMPUS, D. F. 1949. Quantitative ecology of the plankton of the western North Atlantic. *Bull. Bingham Oc. Coll.*, vol. 12, p. 1–169.

ROGER, C. 1973a. Recherches sur la situation trophique d'un groupe d'organismes pélagiques (Euphausiacea): V. Relations avec les thons. *Mar. Biol.*, vol. 19, p. 61–5.

——. 1973b. Recherches sur la situation trophique d'un groupe d'organisms pélagiques (Euphausiacea): II. Comportements nutritionnels. *Mar. Biol.*, vol. 18, p. 317–20.

——. 1975. Rythmes nutritionnels et organisation trophique d'une population de crustaces pélagiques (Euphausiacea). *Mar. Biol.*, vol. 32, p. 365–78.

ROGER, C.; GRANDPERRIN, R. 1976. Pelagic food-webs in the tropical Pacific. *Limnol. Oceanogr.*, vol. 21, p. 731–5.

ROSS, G. J. S. 1970. The efficient use of function minimization in nonlinear maximum likelihood estimation. *Appl. Stat.*, vol. 19, p. 205–21.

ROUGHGARDEN, J. 1974. Population dynamics in a spatially varying environment: how population size 'tracks' spatial variations in carrying capacity. *Am. Nat.*, vol. 108, p. 649–64.

——. 1975a. Population dynamics in a stochastic environment: spectral theory for the linearized N-species Lotka-Volterra competition equations. *Theor. Popul. Biol.*, vol. 7, p. 1–12.

——. 1975b. A simple model for population dynamics in stochastic environments. *Am. Nat.*, vol. 109, p. 713–36.

——. 1977. Patchiness in the spatial distribution of a population caused by stochastic fluctuations in resources. *Oikos*, vol. 29, p. 52–9.

RUSSEL, F. S.; SOUTHWARD, A. J.; BOALCH, G. T.; BUTLER, E. I. 1971. Changes in biological conditions in the English Channel off Plymouth during the last half century. *Nature* (Lond.), vol. 234, p. 468–70.

RYLAND, J. S. 1964. The feeding of plaice and sand-eel larvae in the southern north sea. *J. Mar. Biol. Assoc. U.K.*, vol. 44, p. 343–64.

SAMEOTO, D. D. 1976. Respiration rates, energy budgets and molting frequencies of three species of euphausids found in the Gulf of St Lawrence. *J. Fish. Res. Brd Can.*, vol. 33, p. 2568–76.

SAND, R. F. 1956. Underwater television in commercial fisheries research. *Proc. Gulf. Caribb. Fish. Inst.*, 8th A. Sess., p. 129–32.

SANDERS, H. L. 1968. Marine benthic diversity: a comparative study. *Am. Nat.*, vol. 102, p. 243–82.

——. 1978. *Florida* oil spill impact on the Buzzards Bay benthic fauna: West Falmouth. *J. Fish. Res. Brd Can.*, vol. 35, p. 717–30.

SANDERS, R. L. 1963. Respiration of the Atlantic cod. *J. Fish. Res. Brd Can.*, vol. 20, p. 373–86.

SCHLICHTER, D. 1978. On the ability of *Anemonia sulcata* (Coelenterata: Anthosozoa) to absorb charged and neutral amino acids simultaneously. *Mar. Biol.*, vol. 45, p. 97–104.

SCHRADER, H. J. 1971. Fecal pellets: role in sedimentation of pelagic diatoms. *Science*, vol. 174, p. 55–7.

SCHWOERBEL, J. 1977. *Einführung in die limnologie*. Stuttgart-New York, Gustav Fischer Verlag. 191 p.

SCOR. 1973. *Monitoring life in the ocean: report of working group 29 on monitoring in biological oceanography*. International Council of Scientific Research Proceedings. 71 p.

SEPERS, A. B. J. 1977. The utilization of dissolved organic compounds in aquatic environments. *Hydrobiologia*, vol. 52, p. 39–54.

SHELBOURNE, J. E. 1962. A predator–prey size relationship for plaice larvae feeding on Oikopleura. *J. Mar. Biol. Assoc. U.K.*, vol. 42, p. 243–52.

SHELDON, R. W.; KERR, S. R. 1972. The population density of monsters in Loch Ness. *Limnol. Oceanogr.*, vol. 17, no. 5, p. 796–8.

SHELDON, R. W.; PARSONS, T. R. 1967. A continuous size spectrum for particulate matter in the sea. *J. Fish. Res. Brd Can.*, vol. 24, p. 909–15.

SHELDON, R. W.; PRAKASH, A.; SUTCLIFFE, W. H., Jr. 1972. The size distribution of particles in the ocean. *Limnol. Oceanogr.*, vol. 17, no. 3, p. 327–40.

SHUSHKINA, E. A. 1972. Production rate and utilization of assimilated food for growth by mysids in the Sea of Japan. *Okeanologiya/Oceanology*, vol. 12, p. 326–37.

SHUSHKINA, E. A.; SOROKIN, YU. I. 1969. Radiocarbon determination of zooplankton production. *Oceanology*, vol. 9, p. 594–601.

SIEBURTH, J. M. 1976. Bacterial substrates and productivity in marine ecosystems. *Annu. Rev. Ecol. Syst.*, vol. 7, p. 259–85.

——. 1977. International Helgoland Symposium. Convener's report on the informal session on biomass and productivity of micro-organisms in planktonic ecosystems. *Helgol. Wiss. Meeresunters.*, vol. 30, p. 697–704.

SIGVALDASON, O. T.; DELUCIA, R. J.; BISWAS, A. K. 1972. The Saint John study. *Proc. Int. Symp. Math. Mod. Tech. Wat. Res. Systems*, p. 576–601.

SILVERT, W.; PLATT, T. 1978. Energy flux in the pelagic ecosystem: a time-dependent equation. *Limnol. Oceanogr.*, vol. 23, p. 813–16.

——; ——. 1980. Dynamic energy flow model of the particle size distribution in pelagic ecosystems. In: W. Charles Kerfoot (ed.), *Evolution and ecology of zooplankton communities*, ch. 66, 754–63. Dartmouth, N.H., The University Press of New England.

SJÖBERG, S. 1977. Are pelagic systems inherently unstable? A model study. *Ecol. Modelling*, vol. 3, p. 17–37.

SJÖBERG, S.; WILMOT, W. 1977. System analysis of a spring phytoplankton bloom in the Baltic. *Contrib. Askoe Lab. Univ. Stockh.*, no. 20. 99 p.

SKYRING, G. W.; CHAMBERS, L. A. 1976. Biological sulphate redaction in carbonate sediments of a coral reef. *Aust. J. Mar. Freshwat. Res.*, vol. 27, p. 595–602.

——; ——. 1979. Sulphate reduction in intertidal sediments. In: J. R. Freney and A. J. Nicholson (eds.), *Sulphur in Australia*. Aust. Acad. Sci. (In press.)

SLAWYK, G.; CULLOS, Y.; AUCLAIR, J. C. 1977. The use of the ^{13}C and ^{15}N isotopes for the simultaneous measurement of carbon and nitrogen turnover rates in marine phytoplankton. *Limnol. Oceanogr.*, vol. 22, no. 5, p. 925–32.

SMITH, C. 1976. When and how much to reproduce: the trade-off between power and efficiency. *Am. Zool.*, vol. 16, p. 763–74.

SMITH, D. F.; BULLEID, N. C.; CAMPBELL, R.; HIGGINS, H. W.; ROWE, F.; TRANTER, D. J.; TRANTER, H. 1979. Marine food-web analysis: an experimental study of demersal zooplankton using isotopically labelled prey species. *Mar. Biol.*, vol. 54, p. 49–59.

SMITH, D. F.; WIEBE, W. J. 1976. Constant release of photosynthate from marine phytoplankton. *Appl. Environ. Microbiol.*, vol. 32, p. 75–9.

——; ——. 1977. Rates of carbon fixation, organic carbon release and translocation in a reef-building foraminifer, *Marginopora vertebralis. Aust. J. Mar. Freshwat. Res.*, vol. 28, p. 311–19.

SMITH, F. E. 1970. Analysis of ecosystems. In: D. E. Reichle (ed.), *Analysis of temperate forest ecosystems. Ecological studies*, vol. 1, p. 7–18. New York, N.Y., Springer Verlag.

SMITH, S. V.; JOKIEL, P. L. 1975. Water composition and biochemical gradients of the Canton Island Lagoon. 2. Budget of phosphorus, nitrogen, carbon dioxide and particulate material. *Mar. Sci. Commun.*, vol. 1, no. 1, p. 75–100.

SMITH, W. 1978. Environmental survey design: a time series approach. *Estuar. Coast. Mar. Sci.*, vol. 6, p. 217–24.

SNOW, N. B.; LEB. WILLIAMS, P. J. 1971. A simple method to determine the O:N ratio of small marine animals. *J. Mar. Biol. Assoc. U.K.*, vol. 51, p. 105–10.

SOLÓRZANO, L. 1969. Determination of ammonia in natural waters by the phenol-hypochlorite method. *Limnol. Oceanogr.*, vol. 14, p. 799–801.

SOONG, T. T. 1973. *Random differential equations in science and engineering.* New York, N.Y., Academic Press. 327 p.

SOROKIN, Y. I. 1969. Bacterial production. Pt I: General methods. In: R. A. Vollenweider (ed.), *A manual on methods for measuring primary production in aquatic environments*, p. 128–46. Oxford, Blackwell Scientific.

SOROKIN, Y. I.; VYSHKVARTSEV, D. E. 1974. Feeding on dissolved organic matter by some marine animals. *Aquaculture*, vol. 2, p. 141–8.

SOURNIA, A. 1976. Oxygen metabolism of a fringing reef in French Polynesia. *Helgol. Wiss. Meeresunters.*, vol. 28, p. 401–10.

SOURNIA, A. (ed.). 1978. *Phytoplankton manual.* Paris, Unesco. 337 p. (Monogr. oceanogr. Methodol. 6.)

SOURNIA, A.; RICARD, M. 1976. Phytoplankton and its contribution to primary productivity in two coral reef areas of French Polynesia. *J. Exp. Mar. Biol. Ecol.*, vol. 21, p. 129–40.

SOUTHWARD, A. J.; SOUTHWARD, E. C. 1970. Observations on the rate of dissolved organic compounds in the nutrition of benthic invertebrates. Experiments on three species of *Pogonophora. Sarsia*, vol. 45, p. 69–95.

STAHL, W. R. 1967. The role of models in theoretical biology. In: F. M. Snell (ed.), *Progress in theoretical biology*, vol. 1, p. 165–218. New York, N.Y., Academic Press.

STEELE, J. H. 1974. *The structure of marine ecosystems.* Cambridge, Mass., Harvard University Press. 128 p.

——. 1977. Ecological modelling of the upper layers. In: E. B. Krauss (ed.), *Modelling and prediction of the upper layers of the ocean*, p. 243–50. Oxford, Pergamon Press.

STEELE, J. H.; FARMER, D. M.; HENDERSON, E. W. 1977. Circulation and temperature structure in large marine enclosure. *J. Fish. Res. Brd Can.*, vol. 34, p. 1095–104.

STEELE, J. H.; FROST, B. W. 1977. The structure of plankton communities. *Phil. Trans. R. Soc. Lond. (B: Biol. Sci.).*, vol. 280, p. 485–534.

STEELE, J. H.: HENDERSON, E. W. 1977. Plankton patches in the northern North Sea. In: J. H. Steele (ed.), *Fisheries mathematics*, p. 1–19. London, Academic Press.

——; ——. 1978. Spatial patterns in North Sea plankton. (Unpubl. MS.)

STEELE, J. H.; MENZEL, D. W. 1962. Conditions for maximum primary production in the mixed layer. *Deep-sea Res. Oceanogr. Abstr.*, vol. 9, p. 39–49.

STEELE, J. H.; MULLIN, M. M. 1977. Zooplankton dynamics. In: E. D. Goldberg, I. N. McCave, J. J. O'Brien and J. H. Steele (eds.), *The sea: ideas and observations on progress in the study of the seas*, vol. 6, p. 857–90. New York, N.Y., John Wiley & Sons.

STEEMANN NIELSEN, E. 1952. The use of radioactive carbon (^{14}C) for measuring organic production in the sea. *J. CIEM*, vol. 18, p. 117–40.

STEPHENS, G. C. 1960. Uptake of glucose from solution by the solitary coral Fungia. *Science*, vol. 131, p. 1532.

——. 1962. Uptake of organic material by aquatic invertebrates. I. Uptake of glucose by the solitary coral, *Fungia scutaria. Biol. Bull.*, vol. 123, p. 648–59.

——. 1963. Uptake of organic material by aquatic invertebrates. II. Accumulation of amino acids by the bamboo worm, *Clymenella torquata. Comp. Biochem. Physiol.*, vol. 10, p. 191–202.

——. 1964. Uptake of organic material by aquatic invertebrates. III. Uptake of glycine by brackish water annelids. *Biol. Bull.*, vol. 126, p. 110–62.

STEPHENS, G. C.; KERR, N. S. 1962. Uptake of phenylalanine by *Tetrahymena pyriformis*. *Nature* (Lond.), vol. 194, p. 1094–5.

STEPHENS, G. C.; SCHINSKE, R. A. 1958. Amino acid uptake in marine invertebrates. *Biol. Bull.*, vol. 115, p. 341–2.

——; ——. 1961. Uptake of amino acids by marine invertebrates. *Limnol. Oceanogr.*, vol. 6, p. 175–81.

STEPHENS, G. C.; VIRKAR, R. A. 1966. Uptake of organic material by aquatic invertebrates. IV. The influence of salinity on the uptake of amino acids by the brittle star, *Ophiactis arenosa*. *Biol. Bull.*, vol. 131, p. 172–85.

STEPHENSON, W.; SEARLES, R. B. 1960. Experimental studies on the ecology of intertidal environments at Heron Island. I. Exclusion of fish from beach rock. *Aust. J. Mar. Freshwat. Res.*, vol. 11, p. 241–67.

STEVENSON, R. A., Jr. 1972. Effect of current and light on feeding behaviour and spatial relationships of *Eupomacentrus partitus*, a plankton feeding reef fish. In: H. E. Winn and B. L. Olla (eds.), *Behaviour of marine animals. Current perspectives in research*, p. 278–302. New York, N.Y., Plenum Press.

STEWART, W. D. P.; FITZGERALD, G. P.; BUNIS, R. H. 1967. *In situ* studies on N_2 fixation using the acetylene reduction technique. *Proc. NAS (U.S.)*, vol. 58, p. 2071–8.

STRICKLAND, J. D. H.; PARSONS, T. R. 1965. A manual of seawater analysis. 2nd ed. *Bull. Fish. Res. Brd Can.*, vol. 125. 203 p.

——; ——. 1972. A practical handbook of seawater analysis. 2nd ed. *Bull. Fish. Res. Brd Can.*, vol. 167. 311 p.

SUNDARAM, T. R.; REHM, R. G. 1973. The seasonal thermal structure of deep temperate lakes. *Tellus*, vol. 25, p. 157–67.

SVERDRUP, H. U. 1953. On conditions for the vernal blooming of phytoplankton. *J. Cons.*, vol. 18, p. 287–95.

TAYLOR, C. G. 1953. Nature of variability in trawl catches. *Bull. U.S. Bir. Fish.*, vol. 54, p. 145–56.

TESTERMAN, J. K. 1972. Accumulation of free fatty acids from seawater by marine invertebrates. *Biol. Bull.*, vol. 142, p. 160–77.

THOMAS, W. H. 1964. An experimental evaluation of the ^{14}C method for measuring phytoplankton production, using cultures of *Dunaliella primolecta* Butcher. *Fish. Bull.*, vol. 63, p. 273–92.

THORP, J. H.; GIBBONS, J. W. 1978. *Energy and environmental stress in aquatic systems: DOE Symp.* (ser. 48). U.S. National Technical Information Service, U.S. Dept of Energy. 854 p. (CONF-771114.)

TOMOVIĆ, R. 1963. *Sensitivity analysis of dynamic systems*. New York, N.Y., McGraw-Hill. 142 p.

TUCK, J. L.; MENZEL, M. T. 1972. The superperiod of the non-linear weighted string (FPU) problem. *Adv. Math.*, vol. 9, p. 399–407.

ULAM, S. M. 1963. Some properties of certain non-linear transformations. In: S. Drobot and P. Viebock (eds.), *Mathematical models in physical sciences*, p. 85–95. Englewood Cliffs, N.J., Prentice-Hall.

ULANOWICZ, R. E. 1972. Mass and energy flow in closed ecosystems. *J. Theor. Biol.*, vol. 34, p. 239–53.

——. 1979a. Prediction, chaos and ecological perspective. In: E. A. Halfon (ed.), *Theoretical systems ecology*, p. 107–17. New York, N.Y., Academic Press.

——. 1979b. Diversity, stability and self organization in ecological communities. *Oecologia*, vol. 43(3), p. 295–8.

——. 1980. An hypothesis on the development of natural communities. *J. Theor. Biol.*, vol. 85, p. 223–45.

ULANOWICZ, R. E.; KEMP, W. M. 1979. Toward canonical trophic aggregations. *Am. Nat.*, vol. 114(6), p. 871–83.

UNESCO. 1968. *Zooplankton sampling*, p. 174. Paris, Unesco. (Monogr. oceanogr. Methodol. 2.)

——. 1974a. Monitoring life in the oceans. *Unesco Tech. Pap. Mar. Sci.*, no. 15, p. 71.

——. 1974b. A review of methods used for quantitative phytoplankton studies. *Unesco Tech. Pap. Mar. Sci.*, no. 18, p. 27.

——. 1977. Marine ecosystem modelling in the Eastern Mediterranean. *Unesco Rep. Mar. Sci.*, no. 1, p. 84.

——. 1977. Marine ecosystem modelling in the Mediterranean. *Unesco Rep. Mar. Sci.*, no. 2, p. 111.

VERNBERG, W. B.; COULL, B. C. 1974. Respiration of an interstitial ciliate and benthic energy relationships. *Oecologia*, vol. 16, p. 259–64.

VINOGRADOV, M. E.; MENSHUTKIN, V. V. 1977. Zooplankton dynamics. In: E. D. Goldberg, I. N. McCave, J. J. O'Brien and J. H. Steele (eds.), *The sea: ideas and observations on progress in the study of the seas*, vol. 6, p. 857–90. New York, N.Y., John Wiley & Sons.

VOGEL, A. (ed.). 1961. Ion-exchange and chromatographic methods in analysis. In: *A textbook of quantitative inorganic analysis*, p. 702–37. London, Longmans, Green.

WADA, E.; TSUJI, T.; SAINO, T.; HATLORI, A. 1977. A simple procedure for mass spectrometric microanalysis of ^{15}N in particulate organic matter with special reference to ^{15}N-tracer experiments. *Anal. Biochem.*, vol. 80, p. 312–18.

WAIDE, J. B.; WEBSTER, J. R. 1976. Engineering systems analysis: applicability to ecosystems. In: B. C. Patten (ed.), *Systems analysis and simulation in ecology*, vol. IV, p. 329–71. New York, N.Y., Academic Press.

WALSH, J. J. 1975. A spatial simulation model of the Peru upwelling ecosystem. *Deep-Sea Res. Oceanogr. Abstr.*, vol. 22, p. 201–36.

——. 1977. A biological sketchbook for an eastern boundary current. In: E. D. Goldberg, I. N. McCave, J. J. O'Brien and J. H. Steele (eds.), *The sea: ideas and observations on progress in the study of the seas*, vol. 6, p. 923–68. New York, N.Y., John Wiley & Sons.

WEBB, K. L.; DU PAUL, W. D.; WIEBE, W.; SOTTILE, W.; JOHANNES, R. E. 1975. Enewetak (Eniwetok) Atoll: aspects of the nitrogen cycle on a coral reef. *Limnol. Oceanogr.*, vol. 20, p. 198–210.

WHEELER, W. B. 1970. Experimental absorption of dieldrin by chlorella. *Agric. Food Chem.*, vol. 18, p. 416–19.

WHITTLE, P. 1962. Topographic correlation, power-law covariance functions, and diffusion. *Biometrika*, vol. 49, p. 305–14.

WICKETT, W. P. 1967. Ekman transport and zooplankton concentration in the north Pacific Ocean. *J. Fish. Res. Brd Can.*, vol. 24, p. 581–94.

WIEBE, P. H.; HULBURT, E. M.; CARPENTER, E. J.; JAHN, A. E.; KNAPP, G. P.; BOYD, S. H.; ORTNER, P. B.; COX, J. L. 1976. Gulf Stream cold core rings: large scale interaction sites for open ocean plankton communities. *Deep-sea Res. Oceanogr. Abstr.*, vol. 23, p. 695–710.

WIEBE, W. J.; JOHANNES, R. E.; WEBB, K. L. 1975. Nitrogen fixation in a coral reef community. *Science*, vol. 188, p. 257–9.

WIEBE, W. J.; SMITH, D. F. 1977a. ^{14}C-labelling of the compounds excreted by phytoplankton for employment as a realistic tracer in secondary productivity measurements. *Microbiol. Ecol.*, vol. 4, p. 1–8.

——. 1977b. Direct measurement of dissolved organic carbon release by phytoplankton and incorporation by microheterotrophs. *Mar. Biol.*, vol. 42, p. 213–23.

WILLIAMS, D. D.; BLACKLY, C. H.; MILLER, R. R. 1952. Determination of trace oxygen in gases. *Anal. Chem.*, vol. 24, p. 1819–21.

WILLIAMS, P. J. le B.; BERMAN, T.; HOLM-HANSEN, O. 1976. Amino acid uptake and respiration by marine heterotrophs. *Mar. Biol.*, vol. 35, p. 41–7.

WINBERG, G. G. 1960. Rate of metabolism and food requirements of fishes. *Fish. Res. Brd Can. Trans.*, ser. 194.

WINKLER, L. W. 1888. Die bestimmung des im wassergelosten sauerstoffes. *Ber. Deut. Chem. Ges.*, vol. 21, p. 2843–54.

WINTER, D. F.; BANSE, K.; ANDERSON, G. C. 1975. The dynamics of phytoplankton blooms in Puget Sound, a fjord in the northwestern United States. *Mar. Biol.*, vol. 29, p. 139–76.

WOOD, E. D. F.; ARMSTRONG, F. A. J.; RICHARDS, F. A. 1967. Determination of nitrate in sea water by cadmium-copper reduction to nitrite. *J. Mar. Biol. Assoc. U.K.*, vol. 47, p. 23–31.

WRIGHT, R. T.; HOBBIE, J. B. 1965. Uptake of organic solutes in lake water. *Limnol. Oceanogr.*, vol. 10, p. 222–8.

——; ——. 1966. Use of glucose and acetate by bacteria and algae in aquatic ecosystems. *Ecology*, vol. 47, p. 447–54.

WROBLEWSKI, J. S. 1977. A model of phytoplankton plume formation during variable Oregon upwelling. *J. Mar. Res.*, vol. 35, p. 357–94.

WROBLEWSKI, J. S.; O'BRIEN, J. J. 1976. A spatial model of phytoplankton. *Mar. Biol.*, vol. 35, p. 161–75.

WYATT, T. 1976. Food chains in the sea. In: D. H. Cushing and J. J. Walsh (eds.), *The ecology of the seas*, p. 341–58. Philadelphia, Pa, W. B. Saunders.

YOUNG, D. 1970. *The distribution of cesium, rubidium and potassium in the quasi-marine ecosystem of the Salton Sea*. San Diego, Calif., University of California. (Ph.D. diss.)

155

Young, D. K.; Bazas, M. A., Young, M. W. 1976. Species densities of macrobenthos associated with seagrass. A field experimental study of predation. *J. Mar. Res.*, vol. 34, p. 577–92.
Zotin, A. I. 1973. A phenomenological theory of development. *Ontogeney*, vol. 4(1), p. 3–10.
Zotina, R. S.; Zotin, A. I. 1978. Differential equations of developmental biology. In: I. Lamprecht and A. I. Zotin (eds.), *Thermodynamics of biological processes*, p. 121–41. Berlin, de Gruyter.

Recommended mathematics readings

For readers lacking background in particular areas of mathematics, we can recommend the following texts:

BELLMAN, R. 1970. *Introduction to matrix analysis*. New York, N.Y., McGraw-Hill. 403 p.

LONG, R. R. 1963. *Engineering science mechanics*. Englewood Cliffs, N.J., Prentice-Hall. (Ch. 9 for dimensional analysis and scaling.) 433 p.

MCELIECE, R. J. 1977. *The theory of information and coding*. Reading, Mass., Addison-Wesley. (Ch. 1 for information indices.) 302 p.

SCHEFFE, H. 1959. *The analysis of variance*. New York, N.Y., John Wiley & Sons. 477 p.

YAN, C. S. 1969. *Introduction to input–output economics*. New York, N.Y., Holt, Reinhardt & Winston. 134 p.

In addition we recommend the following texts already cited in the bibliography:

Goel and Richter-Dyn, 1974; Hearon, 1963; Jenkins and Watts, 1968; Patten, 1971; Soong, 1973; and Tomović, 1963.